Games: Purpose and Potential in Education

Christopher Thomas Miller
Editor

Games: Purpose and Potential in Education

Springer

Editor
Dr. Christopher Thomas Miller
Morehead State University
Dept. Curriculum & Instruction
401K Ginger Hall
Morehead, KY 40351
USA
c.miller@morehead-st.edu

ISBN: 978-0-387-09774-9   e-ISBN: 978-0-387-09775-6

© Springer Science+Business Media, LLC 2008
All rights reserved. This work may not be translated or copied in whole or in part without the written permission of the publisher (Springer Science+Business Media, LLC, 233 Spring Street, New York, NY 10013, USA), except for brief excerpts in connection with reviews or scholarly analysis. Use in connection with any form of information storage and retrieval, electronic adaptation, computer software, or by similar or dissimilar methodology now know or hereafter developed is forbidden.
The use in this publication of trade names, trademarks, service marks, and similar terms, even if the are not identified as such, is not to be taken as an expression of opinion as to whether or not they are subject to proprietary rights.

Printed on acid-free paper

springer.com

Dedicated to Laura for her continuous support
and to Sarah and Keegan, my little gamers

Contents

Contents vii

Contributors xiii

List of Abbreviations xv

Preface xvii
    Acknowledgements xvii

Chapter 1 Once Upon a Game: Rediscovering the Roots of Games in Education 1
    Luca Botturi and Christian Sebastian Loh

    1.1 Introduction 1
        1.1.1 Play is Not the Opposite of Work 2
        1.1.2 "What, Then, is Game?" 3
        1.1.3 Chapter Structure 5
    1.2 Games as an Educational Technology 5
        1.2.1 Games and Playing: Theoretical Frameworks 6
        1.2.2 Formal approach: What does a game look like? 7
        1.2.3 Substantial Approach: What is a Game? 8
        1.2.4 Summary: Theoretical Frameworks 12
    1.3 Game and Play: What the Words Say 12
        1.3.1 Play, Play, and Play 13
    1.4 Otium vs. Negotium 16
        1.4.1 School and Teachers 17
    1.5 Conclusion 18
        1.5.1 Education as Game 19
    1.6 Acknowledgments 20
    1.7 References 20

Chapter 2 Learning by Designing Homemade PowerPoint Games 23
    Lloyd Rieber, Michael Barbour, Gretchen Thomas, and Dawn Rauscher

    2.1 Introduction 23
    2.2 Game Design: The Other Use of Gaming in Education 25
    2.3 Constructionism as a Rationale for Gaming in Education 27
    2.4 Game Design as a Route to Play and Flow 29
        2.4.1 Homemade PowerPoint Games: How Do They Work? 31
        2.4.2 What does a homemade PowerPoint game look like? 32

  2.4.3 Homemade PowerPoint Games are an Example of Appropriate Technology...34
 2.5 Homemade PowerPoint Games in Use: K-12 Education..........................34
  2.5.1 Stage 1: Student Orientation 35
  2.5.2 Stage 2: Student Game Design — Scaffolding by the Teacher..........36
  2.5.3 Stage 3: Student Game Design — Student Story Writing..................36
  2.5.4 Stage 4: Student Game Design — Developing an Early Prototype....37
  2.5.5 Stage 5: Student Game Design — Refining the Prototype................37
  2.5.6 Stage 6: Student Game Design — Iterative Cycle.............................38
 2.6 Conclusions..39
 2.7 References...40

Chapter 3 Video Games, Learning, and "Content"..43
James Paul Gee

 3.1 Experience and Learning..43
 3.2 A Piece of Research: Action, Simulation, and Reading.............................45
 3.3 Social Identity and Learning..45
 3.4 Game Design 46
 3.5 The Situated Learning Matrix..47
 3.6 Clearing Up Possible Misconceptions...51
 3.7 References...51

Chapter 4 Fair Game: Gender Differences in Educational Games.................55
Kimberely Fletcher Nettleton

 4.1 Introduction..55
 4.2 Gender Differences in Toys and Games..56
 4.3 Fantasy Play, Simulations, and Games..58
 4.4 The Playground...61
 4.5 Games 62
 4.6 Gender and Education 65
 4.7 Games in the Classroom 66
 4.8 Game Design for Education 69
 4.9 References...71

Chapter 5 Video Game Pedagogy: Good Games = Good Pedagogy 73
Katrin Becker

 5.1 Introduction...74
 5.2 Studying the Masters, and the Scholars..75
 5.3 Connecting the Dots 76
 5.4 On Choosing Games for Study..77
  5.4.1 The Chosen Ones 78

5.5 Game Elements 82
5.6 Learning and Instructional Design Theories and Models 84
5.7 The Classics Revisited 85
    5.7.1 Gagné's Nine Events of Instruction 85
    5.7.2 Reigeluth's Elaboration Theory 92
    5.7.3 Merrill's First Principles of Instruction 100
5.8 New Frontiers 104
    5.8.1 Activity theory 105
    5.8.2 Constructivist Learning Environments 109
    5.8.3 Problem-based learning 114
5.9 Digital Games Are Special (Educational Technologies)......................... 119
5.10 Conclusion 122
5.11 References 122

Chapter 6 Applying Pedagogy during Game Development to Enhance Game-Based Learning 127
Atsusi Hirumi and Christopher Stapleton

    6.1 Applying Pedagogy during Game Development to Enhance
    Game-Based Learning 128
    6.2 Levels of Design and Application 129
    6.3 Fundamental Components of Interactive Entertainment 131
    6.4 Applying Pedagogy during the Game Development Process 132
        6.4.1 The Concept Development Phase 134
        6.4.2 Pre-Production Phase 140
        6.4.3 Production Phase 158
    6.5 Conclusion 159
    6.6 References 160

Chapter 7 Video Games and Teacher Development: Bridging the Gap in the Classroom 163
Elizabeth Simpson and Susan Stansberry

    7.1 Introduction 163
    7.2 Use of Video Games in the Classroom 167
        7.2.1 There is always an answer 167
        7.2.2 Nothing is impossible 167
        7.2.3 Trial and error 167
        7.2.4 Competition and collaboration 168
        7.2.5 Roles are clear 168
        7.2.6 Gamers are autonomous 168
        7.2.7 Gamers dominate their culture 168
    7.3 Teachers' Barriers to Video Games in the Classroom 169
        7.3.1 Accountability 169

7.3.2 Research-based tools and methodologies 170
7.3.3 Administrative Support for Innovation 171
7.3.4 Professional Collaboration 172
7.3.5 Teacher preparedness 172
7.3.6 Need to scaffold new methodologies to existing practice 173
7.4 Strategies for Successfully Integrating Video Games in the Classroom-Bridging the Gap 173
7.5 References 183

Chapter 8 Confronting the Dark Side of Video Games 185
Christian Sebastian Loh

8.1 Introduction 185
    8.1.1 Unique Features of Video Games 186
    8.1.2 The Holodeck Experience 187
8.2 Video Game Playing 188
    8.2.1 Practice Makes Perfect? 188
    8.2.2 Deliberate Practice 188
    8.2.3 Immediate Feedback 190
    8.2.4 The Source of the Feedback 191
8.3 The Rising Controversy 192
    8.3.1 The Debate about Violent Video Games 192
    8.3.2 'R' is for… 195
8.4 In A Galaxy Far, Far, Away… 202
    8.4.1 Marketing to Children 203
8.5 Rating Video Games 205
    8.5.1 A Failing Scheme 206
    8.5.2 Independent Rating System 208
8.6 Conclusion 210
    8.6.1 Epilogue: Which Side Are You On? 211
8.7 References 212

Chapter 9 Blogging the Future from Multiple Perspectives: Current Problems and Future Potentials for Educational Games 219
Christopher T. Miller, Christian Sebastian Loh, Katrin Becker, Luca Botturi, Michael Barbour, Kimberely Fletcher Nettleton, Atsusi Hirumi, Lloyd Rieber, and Elizabeth Simpson

9.1 Introduction 219
9.2 Importance of Connecting Games and the Instructional Design Field 220
    9.2.1 Atsusi 220
    9.2.2 Kimberely 220
    9.2.3 Luca 221
    9.2.4 Lloyd 221

  9.2.5 Katrin 222
  9.2.6 Luca 223
  9.2.7 Christopher 223
  9.2.8 Sebastian 224
9.3 Problems with Games and Education 224
  9.3.1 Katrin 225
  9.3.2 Michael 225
  9.3.3 Kimberely 226
  9.3.4 Sebastian 227
  9.3.5 Luca 228
9.4 Understanding the Benefits of Games 229
  9.4.1 Luca 229
  9.4.2 Kimberely 230
  9.4.3 Katrin 231
  9.4.4 Michael 231
  9.4.5 Sebastian 232
9.5 Integrating Games into Teacher Preparation and Education 232
  9.5.1 Katrin 233
  9.5.2 Elizabeth 233
  9.5.3 Luca 234
  9.5.4 Michael 235
  9.5.5 Kimberely 236
  9.5.6 Michael 236
  9.5.7 Sebastian 236
9.6 What are the Cultural Implications of Increasing the Use of Games in Education? 237
  9.6.1 Katrin 237
  9.6.2 Luca 238
  9.6.3 Michael 239
  9.6.4 Kimberely 239
  9.6.5 Sebastian 239
9.7 How Can Assessment be Conducted When Using Games? 240
  9.7.1 Luca 240
  9.7.2 Katrin 240
  9.7.3 Michael 241
  9.7.4 Kimberely 242
  9.7.5 Sebastian 242
9.8 What Are the Future Potentials for Games in Education? 243
  9.8.1 Michael 243
  9.8.2 Katrin 244
  9.8.3 Katrin 245

  9.8.4 Kimberely 245
  9.8.5 Sebastian 246
 9.9 Conclusion 247
 9.10 References... 249

Biographies 251
 Editor Biography 251
 Author Biographies.. 251

Index 257

Contributors

Michael Barbour
Instructional Technology Department
Wayne State University
Detroit, Michigan 48202
mkbarbour@gmail.com

Katrin Becker
University of Calgary
Alberta, Canada
becker@minkhollow.ca

Luca Botturi
Università della Svizzera italiana
Lugano, Switzerland
luca.botturi@lu.unisi.ch

James Paul Gee
Division of Curriculum and Instruction
Arizona State University
Phoenix, AZ 85004
James.Gee@asu.edu

Atsusi Hirumi
Instructional Technology
University of Central Florida
Orlando, Florida, 32816
hirumi@mail.ucf.edu

Christian Sebastian Loh
Department of Curriculum & Instruction
Southern Illinois University Carbondale
Carbondale, IL 62901
csloh@siu.edu

Christopher Thomas Miller
Department of Curriculum and Instruction
Morehead State University
Morehead, Kentucky 40351
c.miller@morehead-st.edu

Kimberely Fletcher Nettleton
Department of Curriculum and Instruction
Morehead State University
Morehead, Kentucky 40351
k.nettleton@morehead-st.edu

Dawn Rauscher
Flathead Valley Community College
Kalispell, Montana 59901
drauscher@gmail.com

Lloyd P. Rieber
Department of Educational Psychology & Instructional Technology
The University of Georgia
Athens, Georgia 30602
lrieber@uga.edu

Susan Stansberry
Educational Technology
Oklahoma State University
Stillwater, OK 74078
susan.stansberry@okstate.edu

Christopher Stapleton
Media Convergence Laboratory
University of Central Florida
Orlando Florida 32826
cstaplet@ist.ucf.edu

Elizabeth S. Simpson
Department of Special Education
University of Wyoming
Laramie, WY 82071
lsimpson@uwyo.edu

Gretchen Thomas
Department of Educational Psychology & Instructional Technology
The University of Georgia
Athens, Georgia 30602
gbthomas@uga.edu

List of Abbreviations

ACM	Association for Computing Machinery
ACWW	Animal Crossing Wild World
AECT	Association for Educational Communications and Technology
APA	American Psychological Association
COTS	commercial off the shelf
DMT	dialogue macrogame theory
DHHS	Department of Health and Human Services
ESRB	Entertainment Software Rating Board
EU	European Union
FAS	Federation of American Scientists
FTC	Federal Trade Commission
GBL	game-based learning
GD	game development
H.U.D.	heads up display
ID	instructional design
ID2	second generation instructional design
iDPA	InterPlay Digital Assistant
ISMF	International Student Media Festival
KoToR	Star Wars: Knights of the Old Republic
L.O.D.	level of detail
MMOG	massive multiplayer online games
MMORPG	massive multiplayer online role-playing game
NCLB	No Child Left Behind
NPC	Non-player character
NWN	Neverwinter Nights
OFB	Office for Film and Broadcasting
P.O.V.	point of view
PBIS	positive behavior intervention supports
PBL	project based learning
PC	personal computer
PLC	professional learning community
PS2	Playstation 2
PW	Phoenix Wright: Ace Attorney
RPG	role-playing game
RtI	response to intervention
SMB	Super Mario Bros.
TESIV	Elder Scrolls IV
UDB	understanding by design
UMPC	ultra-mobile personal computers
VGBL	video game-based learning
WWW	World Wide Web

Preface

This book provides a general introduction to the use of games in education. Games in education have become an increasingly hot topic in the field of educational technology in the beginnings of the 21st century and will continue to grow as the technologies powering games evolve. It is important that in the field of education that we better understand the purpose and potentials of games in education. It is with this in mind that this book originated in an airport on the way to the 2006 Association for Educational Communications (AECT) annual convention. This book was designed individuals in the field of education to better understand several of the main topics related to games in education. The topics in this book vary from exploring the concept of games and its connection to education, the design of games in education, gender, identity, pedagogy, instructional design, games and teacher development, and the dark side of games. Additionally, several of the contributors of this book participated in a year-long blog-based discussion about additional problems and future potentials for games in education. I hope that the book will be informative, engaging, thought provoking, and even, dare I say it maybe fun.

Acknowledgements

I wish to express my appreciation to colleagues who contributed chapters to this exploration of games in education. I'd also like to thank my colleagues Dr. James Knoll, Dr. Wayne Willis, and Mr. William Cole, who helped mentor me through the editing process and listened to all of my questions. The support of Marie M Sheldon and Kristina Wiggins-Coppola at Springer for their support in moving this project to its completion are appreciated. Finally, my thanks to AECT and especially Dr. Phillip Harris, who helped make the connections to Springer Publishing.

Chapter 1
Once Upon a Game:

Rediscovering the Roots of Games in Education

Luca Botturi and Christian Sebastian Loh

Abstract

In view of the recent interest in using videogames for learning, many teachers and parents have begun to question the place of videogames in the classrooms. In this chapter, we attempt to explore the very idea of playing and learning by trying to rediscover the hidden meanings in usual words, like "game," "play," "school" and "education" through a lexical and conceptual analysis within the Western culture, roaming among ancient and modern languages. It is through the rediscovery of our roots that we as educators can be better informed to either embrace or discard the call to integrate play into education for game-based instruction.

1.1 Introduction

"Let my playing be my learning, and my learning be my playing."
- Johan Huizinga

While it took several millennia for games to evolve from being played in a sandbox to a virtual video world; it has taken only a couple of decades for video games to progress from mere moving dot and lines (e.g. *Pong*) to 3-Dimensional graphical avatars playable on the Internet (e.g. *World of Warcarft*). At one time, particularly in the 1970s, the term *video games* meant "games played in a video arcade." However, in today's context (and for this chapter), the term is used broadly to include all digital games playable on a device with video screen, which would include computers, game consoles, cellular phones and mobile devices.

In some sectors, including education, business, military, healthcare, and government (Michael & Chen, 2006), the term *serious games* was used to distinguish videogames that are created for training and instruction from those developed for entertainment purposes. Although the concept for serious games (and for that matter, the use of games in classrooms) is not new, what's new is the *media* of video games. As researchers and educators began to delve more into videogames and the complex learning dynamics that take place during game play, it is important for the educational community to clarify what game means in order to facilitate clear dialog with other learning domains and the videogame industry.

1.1.1 Play is Not the Opposite of Work

Even though teachers are no strangers to using games (e.g. board games, card games, and role-playing games) in the classrooms, the primary reason for using them has always been for learning and not for entertainment. When used within a classroom setting, games functioned as a teaching aide in helping to explain or reinforce a learning concept. Sometimes a complex scenario not easily understood through reading alone may be acted out through games (Van Ments, 1999). Therefore, there appeared to be a mismatch: while games were mostly played for the element of *fun* (Koster, 2005), teachers seldom use games for this reason.

Perhaps the reason was today's performance-based curriculum, or it could have been the post-industrialized society. Somehow, somewhere, a viewpoint was born: going to school becomes the *job* of the youth, for their preparation to enter the world of the adults. Consequently, one tends to say that going to school is productive, while playing is not, as this corresponds to the fact that salary is paid based on how much time a person spent in working, not in playing. Hence, it is no surprise that the education system would inherit the same belief to uphold *work* and put down *play*.

While the parallel between school and work is true in the sense that "school is a duty as a job is," it can lead to a misunderstanding when it comes to the nature of what teaching and learning are. Rieber (1996) points out that the opposite of *work* is *leisure*, not *play*. Yet, because *play* could detract one from *work*, it is often regarded as the opposite of *work*. As such, (game) *play* has been relegated to personal time outside of work, and must not come to interfere with real *work,* such as classroom teaching and learning. This belief is also evidenced in the so-called *extra-curricular* (or extramural) activities of schools that are comprised mainly of *play* such as band, chess, dance, and sports.

When computer technology was first introduced into the classrooms some reactions were purely technophobic (Rosen & Weil, 1995). As feedback from our graduate students (*digital native*[1] in-service teachers) revealed, some teachers may be reacting to the new technology in a *ludophobic* manner. They chided the digital-native teachers, who let students play video games in class, as "not doing *real work* during class time." These non-native teachers were also worried that their

---

[1] Digital Natives: so named because they are "native speakers" of the digital language of computers, video games and the Internet — loosely, those born after the 1980s.

own students may begin demanding video games due to the precedent set by other digital-native teachers.

The view that "*schools* have more to do with *playing* than *working*" is simply at odds to traditional thinking. But what is *traditional thinking*? According to Beck and Wade (2004), older non-gamers (aged 32 and above) have problems accepting that videogame playing can be "purposeful and serious." This is contrary to the next generation who grew up with ubiquitous access to digital media, they simply accept that "play *is* work" (Prensky, 2000). As they come of age, traditional thinking may well become one that embraces play as work, and work as play.

If videogames are to become useful for learning, it is imperative for educators to understand *what videogames are*. If they do not, they will either feel threatened by the new technology, or commit enthusiastically to it without any sound reason. This is actually the topic of this chapter. Also, educators will need to carefully consider the merits and potentials of videogames for serious learning, and, lastly, identify suitable games from thousands of available titles for use in the classroom.

But first of all, are we talking about videogames, or games? If we follow the thinking path long enough, we will no doubt arrived at the same conclusion as Smuts (2006), "In order to define *video game*, one must confront the problem of defining *game* itself." The issue of using videogame in classroom learning is intertwined with that of using games in the classroom; for the two problems are, in fact, one.

1.1.2 "What, Then, is Game?"

Educators are not the only one who is perplexed by the lack of a well-defined term. Film writer and game designer, Lee Sheldon (2004), wrote:

> "One problem we come across when attempting to discuss games is our lack of a common vocabulary. We've borrowed terms from other media and then changed their definitions… [and] made up our own words." (p. xiii)

Because of our different world view on games, and because educators are beginning to warm up to videogames as a legitimate field of study (Hill, 2005); it is essential for a shared perspective and understanding to be established between educators and professionals from other disciplines in order to facilitate clear dialogues. We need to develop a more specific definition of what it is to be a video game. There have been very few attempts to define video game, and none of them have been successful. It is perhaps more vital for researchers in the education arena to first have a common agreement of *what games are* before trying to render support to other educators in integrating videogames into classroom practices.

Outside the classroom, playing is one of the fundamental human activities, one of the first that human children develop together with talking, toddling, and relating to others. The experience of playing belongs to the fundamental palette of human experience (Gadamer, 1965), and it is probably as old as human beings are – or even as intelligent life is.

On the other hand, playing has a special feature with respect to other basic human activities, such as eating or sleeping: when we eat or sleep, we eat or sleep. When we play, we can (pretend to) eat or sleep, and eating and sleeping become parts of a playful dimension. Children can decide to play in every situation and with any object, which becomes a toy. For example, when children "play house," they are pretending (playing) to perform daily chores such as cooking and cleaning. When little girls play "mommy", they are imitating (playing) what mothers do to their babies.

Thus, playing is not something that we do distinctly apart from daily life. It is a modality of doing things, a *mode* of human experience, a sort of envelope of what we do that give a specific different hue to the activities that we perform. This mode of experience is natural to children, while it is more difficult to adults (Gadamer, 1965).

What is a game then? The answer cannot come in a few lines, and the whole chapter is a partial effort to provide some initial insight. Yet from this perspective a game is a structured set of rules that create a space (the magic circle described by Salen & Zimmerman, 2004) in which the playing mode of experience is possible to adults. Game-playing is then a specific activity such as eating and sleeping, and in this sense it is possible to distinguish *playing*, as a natural mode of experience, from *game* and *gameplay*, a culture-based activity.

St. Augustine of Ippona noted that while people have no doubt about what *time* is in their everyday life, its definition eludes them: when asked to provide one, the concept seems to get blurred. This is true of all the most familiar experiences of human life, like *love, friendship,* etc. Defining the concept of *game* can be equally elusive. We all probably have a quite clear idea of what a game is based on our experience of playing games, maybe as kids, and have a number of good examples. But how would we define it? The work of Salen and Zimmermann (2003) collects 15 different definitions from the literature, and testifies that there is no easy clear-cut and agreed-upon solution.

One good way to get a new insight into a familiar experience is to reflect on the words that we use to describe that experience. Languages in fact preserve a dense stratification of meaning that we easily overlook or even forget in everyday speech. Readers should be warned that working with concepts and languages is tricky, because they embody the very structure of a culture, and are consequently strongly culture-dependent. For this reason we decided to limit the scope of the analysis to (a part of) Western cultures, in the awareness that while additional analysis outside the Western *Weltanschauung* (literally, *view of the world*) would provide additional extremely interesting elements, this is a sensible starting point for the issues at stake. It is through the rediscovery of our roots that we as educators can get a better understanding to actually improve our educational practices.

1.1.3 Chapter Structure

We would like to propose a challenging thought: What if schools were originally made for the purpose of game and play? In order to support our claim, we will explore the concept for *playing* and *games* using a lexical and conceptual analysis of common words used in everyday lives, such as "game," "play," "school" and "education," to rediscover any hidden meanings in their roots. These words describe basic human experiences, and bring multiple layers of meaning that can provide unexpected insights in the nature of education and playing and in the connection among them.

Since *game* is a human activity and not a specific subject matter (Bittanti, 2004), the nature of this inquiry is decidedly interdisciplinary.

Readers should note that we have deliberately ignored the literature on game studies and game design research for this study, in favor of a strict conceptual and lexical analysis.

1.2 Games as an Educational Technology

Even if only recently it has become a hot topic, games have always been parts of teachers' array of teaching techniques. Education is often the first benchmark for exploring the potential of new technologies (Cantoni & di Blas, 2006) and video games can be considered as a type of educational technology. In the following section, we will examine the views of educators and scholars in the field of educational and instructional technologies on games and game playing.

Having consulted a number of standard instructional technology texts and reference books at both the graduate and undergraduate levels (Alessi & Trollip, 2001; Anglin, 1995; Heinich, Molenda, Russell, & Smaldino, 1999; Jonassen, 1996, 2004; Lever-Duffy, McDonald, & Mizell, 2003; Maddux, Johnson, & Willis, 1992; Smith & Ragan, 1999), we found that very few of them considered *game* to be an instructional resource, technique, or tool. In the rare instances where game was mentioned, it was regarded as a motivational activity to supplementary learning in the classroom. The readers (mostly pre-service teachers) were cautioned against using game for *play* in the classrooms. For example, Heinich and colleagues (1999) referred to games as "activities" in which "participants follow prescribed rules that differ from those of real life as they strive to attain a challenging goal." Gredler (1996, 2004) defined games as "competitive exercises" in which "the objective is to win and players must apply subject matter or other relevant knowledge in an effort to advance in the exercise and win." While Alessi and Trollip (2001) not only placed games under a chapter entitled "Drills" but further suggested "embedding a drill into a game activity."

Firstly, it seemed unlikely for word such as "activities" and "competitive exercises", if found printed on box-covers of video games, to generate any interest from gamers – or from young students. Secondly, with respect to the latest developments of the research on games such definitions sound nowadays rather simplistic. Thirdly, people enjoyed playing games because games are fun, and this is what makes them learn. If educators view games merely as *activities* and *competitive exercises*, they will use them (or design them) as such, hence, jeopardizing the real potential of games.

Based on the *ad hoc* textbook survey, it would seem that *games* were either ignored (or omitted) by educators, or were (mis-)presented as competing factors to classroom learning. In short, the usefulness and potentials of games were downplayed. However, before we criticize these authors for not seeing *games* for what they are worth, is it possible that such unexciting definitions were merely a ploy to pass-off games as *acceptable* learning exercises in yesterday's "anti-play" classrooms? One must not forget that, at one time, *games* were not an acceptable form of learning in schools, at all (Rieber, Smith, & Noah, 1998). So perhaps, some of these forward thinking educators were actually trying to cast *game playing* as exercises so that it may slip by the "game police!"

Whatever the reasons, it sufficed to note that there is an urgent need to update these textbooks both to correct the perspectives of teachers on the potentials of video games for learning and to line up with the digital natives' understanding of games. Even as educators search for a solid ground upon which to build good practices for using games in learning, school teachers need to recalibrate their belief systems to accept learning and fun can coexist in the media of video games. Not only is *play* not an extra-curricular activity, it ought to be encouraged during class time to facilitate experiential learning. In fact, as we will show in later sections, *school* was created originally for the purpose of *play* — at least according to the Ancient Greek language, as we will discuss below.

1.2.1 Games and Playing: Theoretical Frameworks

The second step extends the perspective developed in the first by exploring the meaning of *game* in disciplines other than education. The review presented here is forcedly a non-exhaustive overview, with illustrative purpose. It is possible to divide the exploration in two approaches.

1. *Formal approach*. Some disciplines used the idea of games as a metaphor to describe formally some complex structures proper of their field. The examples reported here come from Economics and Argumentation Theory.
2. *Substantial approach*. Other disciplines tried to investigate the nature of playing and games, and of ludic experience in general. The examples reported here come from Semiotics and Philosophy.

1.2.2 Formal approach: What does a game look like?

1.2.2.1 Game Theory

In about the same period when Wittgenstein (1961) was developing his theory of *Sprachspiele* (literally, *language games*), Nicholas Von Neumann gave birth to game theory (Von Neumann & Morgenstern, 1953), a crossover field between computer science and economics. The game theory, which is a theory of interaction that models complex social interactions (among which also some games which are often used as examples), was later expanded and formalized by John Nash[2] (1951), who won a Nobel Prize in 1984. Games are subdivided as collaborative and competitive games, in which *players* behave as *agents* who follow rules (of the game) and move in turns, in search of some expected *payoff* (reward) for their efforts. Game theory is commonly applied in the representation of conflicts or market dynamics.

Within game theory, the structural elements of a game exist as rules, turns, collaboration and competition, where winning, or fun, is modeled as numerical payoff. Game theory tries to explain *how playing (a game) works*, and defines games as an interactive process striving toward a payoff. Because game theory provides a phenomenal description of a game (i.e., what happens during the game?) without investigating the meaning of the game (i.e., why do we play?), we will label its approach as *functional* approach.

1.2.2.2 Dialogue Macrogame Theory

A similar vision can be found in a branch of linguistics and argumentation theory. Argumentation and dialogue are forms of logical discourse that take place between two or more parties, based on specific logical reasoning and premises. Mann (2002) developed the dialogue macrogame theory (DMT; successor to the dialogue game theory), a model to represent and analyze real verbal dialogic interactions. A dialogue is described as a joint activity of two or more partners that share a goal, and each utterance in the dialogue is analyzed as a *turn*, or *move* within the macrogame. Conventions, courtesy, and the production of meaning are the rules that each game should follow, and breaking then generates a conflict or nonsense.

During the interaction a partner can bid a game, for example *information seeking* (by asking a question), that the others can accept (answering the question, or

---

[2] A visual example of Russell Crowe as John Nash explaining non-cooperative games can be seen in the "bar application" of game theory in the movie A Beautiful Mind (Howard, 2001, 0h18'50"-0h20'50").

asking for classification) and conclude (providing the required information) or refuse (e.g., by changing subject). Each verbal exchange in the dialogue is therefore interpreted as a move of a game within the dialogue itself, which becomes a sequence of games, which in their turn can include other games. Walton (1984) developed a taxonomy of dialogue games according to the kind of shared goal. For example, a mediation game has the goal of settling a conflict in a way that can be accepted to both parties. In this situation the mediator covers a specific role, which includes the possibility to threaten as a particular move in the game; this is not allowed to the conflicting parties, otherwise the whole mediation would fail. Also, threatening is never allowed in a scientific communication game, whose goal is the generation of new knowledge.

Both game theory and the dialogue macrogame theory exploit the formal structure of games in order to define metaphorically useful concepts for modeling complex interactions: rule systems, turns, roles, allowed moves, goals.

IMPLICATION #1: Games are Interactions

In this functional approach, *playing* is an interaction among players. In this respect, a person who is playing with a videogame (standalone) is a limit case in which the game system (the computer partner) is so advanced that it is able to sustain continual interactions with the human player. While the dimension of the videogame playing phenomenon makes it deserve the attention of the research community, it is probably not the best starting point for a game-based, learning investigation. Some notable exceptions include (a) multiplayer videogames (c.f. Heliö, 2004), (b) online games played with human partner(s), and (c) single-player games that provide support for intense social interaction (Gee, 2003).

1.2.3 Substantial Approach: What is a Game?

Some disciplines have wondered about the nature and experience of playing, by questioning its real essence: What is a game? What is playing? We call this a *substantial* approach, which is indeed very different from the *functional* approach presented above. To our purposes, two interesting contributions come from Semiotics and Philosophy. We will navigate through these vast disciplines through the works of selected authors, including Huizinga, Gee and Callois for semiotics, Heraclitus, St. Thomas Aquinas and Schiller for philosophy.

1.2.3.1 Semiotics

The reference point for Semiotics is undoubtedly *Homo Ludens* (Huizinga, 1980). Huizinga describes playing as an experience characterized by several features:

1. First and foremost, a game is a voluntary activity. If the game is not voluntary, then the player will find himself in Michael Douglas' situation in the movie called *The Game* (Fincher, 1997): an apparently meaningless sequence of threatening events.
2. A game exists only within the boundary of defined time and space — when these boundaries are violated, one exits the *game* and enters a nightmare much like that depicted in the movie, *Jumanji* (Johnston, 1995). The delimitations in time and space define a sort of separation ordered by special rules that create an environment different from ordinary life. Huizinga described this situation as a magic circle between reality and the game's fictional world.
3. A game always has an end in itself. A person could play the game for playing's sake, or for fun; but not for something else (a special case being: gambling, which will be discussed later).
4. Once a person begins playing a game, he must commit to playing it to its end, and flow along with the tensions within the game to finally resolve in joy (fun).

Again, the important element here is the concept of *magic circle*: a game must have clear boundaries, both in space and in time. Sports mark their space with lines on the field, children with free spots or by labeling, "this is the office". This leads to the second implication.

IMPLICATION #2: Games are Delimited

Introducing a game in the classroom means creating a sort of *free zone* in which only the game exists, without any influence of extrinsic elements such as evaluation, grades, mandatory work, or other. Playing requires conditions, and this is probably the most challenging issue in the use of games in formal education.

The second key element emphasized by Huizinga is freedom: no one can be compelled to play – or this will not be playing any more, at most just following the rules of a game. This will be the topic mostly analyzed by Philosophy in the following paragraphs, and lead to our third implication.

IMPLICATION #3: Play is Voluntary

Each player must deliberately choose to play, and take it seriously. Considering that being obliged to play is not playing, using games in the classroom can be difficult. Also, in adult learning, taking a game seriously can sound like a joke and even have a counter-effect.

Following the same line, Gee (2003) defined games as a semiotic environment, i.e., not simply an activity, but a structured system of signs, a culture or a world, in which the players develop specific perceptual, understanding and action abilities. A semiotic environment is a working space for reflection on cultural models, values, and identities[3]. For example, playing cards is also learning to become a player, and a certain kind of player (trustworthy, sneaker, sympathetic, etc.). At the same time this requires observing oneself developing a new identity.

So 'playing' involves taking a risk, even a risk that concerns one's identity. Players should be aware and ready for that, or they will not play. This requires trust in who proposes a game, maybe the teacher, and at the same time emphasizes a clear distinction between *instruction* as teaching and learning skills and *education*, where identity and personal growth is at stake.

Also, Gee (2003) believes that (video)games can promote social interaction within (e.g. by joining an online multiplayer game) as well as outside the game world (by joining a game social group). Therefore, games are experiences that foster active and reflective learning by: (a) enhancing the development of the ability to read one's own experience, (b) joining affinity groups, and (c) problem solving in a critical and reflective way.

IMPLICATION #4: Games Play With Identity

Once players enter the "magic circle" of a game, they must respect its rules. Playing along with other players means accepting the game world's reality and negotiating one's identities within the game world as if it is real. Players should be ready for this "reality/identity change." While this type of negotiation may come naturally for children (it's just another make-belief), it can be difficult for adult education.

A final contribution from Semiotics, which is paramount both to educators and game designers, comes from Callois (2001, 1981), which identified four types of games. Actually, these can be seen as four dimensions of playing (or generations of fun), which are represented in different degrees in actual games.

1. *Agon* is the game of competition, such as sports;
2. *Alea* is the game based on chance and risk-taking, such as in gambling;
3. *Mimicry* is the game of make-believe, such as adventure games, or childish make-believe games.
4. *Ilinx* is breath-taking games, in which the challenge to the senses and the overcoming of fear generates pleasure, such as in bungee jumping.

Each game develops from a sort of texture of these elements, which leads to develop the pleasure in games. For example, a pleasant afternoon can be spent rafting a mountain river and challenging one's fear (*ilinx*) and at the same time taking the role of "ship's admiral" (*mimicry*) and racing with other teams (*agon*). This brings us to the fifth implication.

---

[3] See Chapter 3, Video Games, Learning, and Content.

> **IMPLICATION #5: Games Are Not All Alike**
>
> Callois' dimensions indicated that there are different dynamics at work in each game. It is sensible to expect that each dimension bear different effects to learning, e.g., different types of learning facts, skills, etc. Also, different players may or even prefer to play the same game differently, which can be problematic when introducing new games to a class. Also, and again this is a methodological hint, videogames do not cover the full array of game types. And finally, different people will have fun with different games – another challenge for educators which have to deal with diverse populations of students.

1.2.3.2 Philosophy

Philosophy is probably the field of study in which games have found the largest space as a topic. Games and playing were first mentioned in Heraclitus' fragments, the earliest documents of ancient Greek philosophy. God was portrayed as a child playing with the human fate, "The Geschick of being, a child that plays, shifting the pawns: the royalty of a child" (Heraclitus' Fr. 52, c.f. Kahn, 1981).

Aristotle (about 350 B.C.) in his *Nicomachean Ethics* considered playing as a necessary activity for re-establishing balance in the tired soul. St. Thomas Aquinas goes further and observes that players play for playing sake, i.e., for fun and for nothing else (Melchiorre, 2006). This is also referred to as an *autotelic* activity: an activity that is self-aimed and has no external end or purpose – indeed, a feature mentioned by Huizinga in *Homo Ludens*. Under this respect playing is comparable to art and spiritual contemplation: art is the creation of formal beauty for no other goal than enjoying beauty itself, and spiritual contemplation is the search of the vision of God for no other end than the joy seeing God. From the Medieval perspective, and in opposition to our modern contraposition of work and play, the autotelic nature of playing elevated it to the level of poetry and praying. This is indeed what playing and learning (not being taught) have in common: they are an activity performed for itself: for fun, and for becoming a better person.

During the Romantic period in 18th century, Schiller (1775, 2004) extended this claim stating that human beings play only when they are human in the purest sense, and they are purely human only when playing. Thus, *playing* becomes the only time when we can disregard what we *must* do, to focus on what we *desired* to do. In this sense the experience of playing is close to that of art, which for the Romantic Movement was the main thrust in exploring the mystery and meaning of human life.

So what is the meaning beneath Heraclitus's verses that describe God playing with the human fate? On the one hand, this indicates that because God is free from any responsibility to other beings: He can do what he pleases with the fate of the world – this is our feeling of cruelty for a God that "plays" with us. More deeply, God moves the world for no other end or reason except for the pleasure He has in doing so. This is indeed a form of free love, which resembles that between parent and child.

Altogether, Philosophy indicates that playing is not a free-time diversion, but a very important human activity. This brings us to the sixth implication.

> **IMPLICATION #6: Play is not Recreation but Re-Creation**
>
> Play is not only a humble diversion from everyday's life, but a deeply human activity close to music, figurative art, and meditation. It requires free choice and the rational acceptance of rules, which include its being limited to specific moments in time. As such, it has the ability to re-create (i.e., restore) a person's soul.

1.2.4 Summary: Theoretical Frameworks

While the small sampling of topics described above is hardly representative of the contributions that each discipline has on games and playing, and still less of what other disciplines have to say about it, we hope it allows educators, researchers, and game designers to draw appropriate implications from them. In addition, it has allowed us to identify how the disciplines approach the topics of *games* and *play*.

The functional approach to games and playing allows researchers and game designers to better understand how *playing* works, and the implications are to designing better and more effective (serious) games. The substantial approach, on the other hand, examines the very essence of *playing*. By investigating the conditions needed and the true meaning behind game playing, educators can obtain new insights on how to integrate games into an educational environment.

1.3 Game and Play: What the Words Say

This section investigates *games* and *play* by exploring the layers of meaning that compose the two words used in five modern Western languages, namely English, Spanish, Italian, German and French, and their Latin roots. Table 1.1 shows the main verbs and nouns in the six languages.

Table 1.1 Lexical Analysis of "Play" (verb) and "Games" (noun) in Western Languages

-	English	German	French	Italian	Spanish	Latin
Play (verb)	Joke/Mock/Tease	Witzen	Jouer	Scherzare	Bromear	Iocari
	Play	Spielen		Giocare	Jugar	Ludere
				Suonare	Tocar	Sonari
				Recitare	Representar	[Ludere]
Game (noun)	Joke	Witz	Plaisanterie	Scherzo	Broma	Iocus
	Game	Spiel	Jeu	Gioco	Juego	Iocus
	Play		Pièce	Recitare	Representación	Ludus
	Toy	Spielzeug	Jouet	Giocattolo	Juguetes	Crepundia/Tricia/Lusus

1.3.1 Play, Play, and Play

The English verb *play* and the corresponding German *spielen* describe the activity of playing at large, which includes three activities that are distinct in Spanish, Italian and Latin, namely:

1. Playing a game (giocare, jugar, iocari);
2. Playing music or a musical instrument (suonare, tocar, sonari);
3. Playing a theater performance (recitare, representar, ludere).

Strangely, even though French is also a Latin language (like Italian and Spanish), the usage of the word *jouer* is similar to the English *play* – this could be due to the historical influence occurred between the two countries. The English word *play* thus comprises a number of artistic – or, as mentioned above, autotelic – activities, emphasizing the deep connection between playing and the arts. Readers should take note that the semantic association between playing and the arts only applies to performing arts (music and theatre) but *not* figurative arts – a point to be explained later by the examination of the lexical root for *play* and *spielen*. By now it is important to point out that different languages and cultures mark different borderlines among these concepts. The connection between games and performing arts, which is intuitive for English and German native speakers, should not be taken for granted at large.

1.3.1.1 Iocus and Ludus

Latin uses two main words for *playing* and *games*, which survives till today in modern Latin languages: *iocari/iocus* and *ludere/ludus* (Usener, 1979). From these words stem (almost) all related verbs in French, Spanish and Italian[4].

The Latin *iocari* means "saying something to induce laughter," e.g., making fun with words like in a pun or a joke. The word probably originates from the Indo-European root *jehan*, which means "to say, pronounce," from which also come the English *yes* and the German *ja*. From *iocus* also comes the German word *juwel*, the Italian word *gioiello* (both meaning *jewel*), and the English *joke* — indicating that having fun is something precious, beautiful and valuable. Again, this is another hint at the strong relationship among games, playing, fun, pleasure, and beauty (such as in the arts), as discussed earlier.

Ludus, on the other hand, denotes the action of physical playing, such as in sports, competitions, institutionalized games (e.g., the Olympic games); The plural, *ludii* was used for public performance events such as theatre and gladiators' show (*ludii gladiatorii*). The word *ludus* will show up again in the next section. Concerning modern languages, both *ludus* (noun) and *ludere* (verb) survive only as derivative forms in unexpected words, such as *illusion* (the experience of being captured in a fictive and false believing), *delusion* (coming out of a fictive believing, with the consequent disappointment), *conclusion* (bring the game to an end), *prelude* (what comes right before the game starts), etc. Interestingly, similar derivatives exist also in German as the suffix – *spielen*. For instance, *anspielen* means to "illude."

The dichotomy between *iocus* and *ludus* brings to mind Callois's dimensions of games: some are make-belief activities (*mimicry* and to a certain extent *alea*), while others require a physical dimension (*agon* and *ilinx*). Hence, the word *play* can be used to mean "playing a large variety of games."

1.3.1.2 Play, and Spielen

Both the German spielen and English play come from a Nordic root, *spil[5], which denotes a happy, dancing movement, such as the one displayed in swordplay, or in the German "Die Hand mit im Spiel haben" (literally, "the hand is playing with something", which means "being busy with something"). Thus, play also involves movement and action, and, as we pointed out above, is associated with performing arts, music and theatre (or dance), but not figurative arts. Notice that the semiotic definition for game fits a theatrical *play* very well: we see a specific set of semi-

---

[4] The exception being the Italian *scherzare* and the Spanish *bromear* (equivalent to the English *mock* or *tease*) which came from more recent vernacular roots.

[5] The * indicates ancient words that are not used in that form any more in a language.

otic codes (dialogue, facial cues, movements, etc.) being played out within a delimited environment (stage and time).

The deep relationship among the different languages is further illustrated by the metaphor: "There is some *play* between the gear's levers and cogs." Interesting, this very metaphor exists almost verbatim in Italian, Spanish, French and German. In this case, *play* is presented as a sort of free movements within a rigid or confined structure. This brings us to the seventh implication.

IMPLICATION #7: Play as Free Movement

As the word *play* implies, a *game* is essentially: a series of free movements within a well-defined (rigid) structure. Potentially, this is how videogames may be used within a formalized structure (e.g., the educational system). However, any revolutionary man becomes a conservative the day after he won: if the free space within a rigid structure is institutionalized, is it still a *free* space?

1.3.1.3 Gamen, and Toy

Finally, the word *game* (and its close relative: *gamble*) comes from the ancient English word *gamen*, which described a meeting or party, or a moment of joy, amusement, sharing and communion. The concept of *game* thus carries with it an intrinsic social dimension, echoing what we already observed from Game Theory and the Dialogue Macrogame Theory: playing is interactions, and having fun requires being with others, like dancing or making music. This brings us to the eighth implication.

IMPLICATION #8: Play is Social

Because *play* is fundamentally a social function, this makes *game* a social event. Remember that even in a standalone videogame session, both the human player and the game console (or computer) are partners in the same social (magic) circle. Massive Multiplayer Online Games are huge social events, and so are weekly game club meetings.

Some readers may have heard the expression that "videogames are complex toys." What, then, is a *toy*? The English *toy* is strictly bound with the word *tool*, and has also connections with several words in other languages, including the Dutch *tiug*, the German *zeug*, and the Swedish *tyg*, which all means gears, or tools. The German word for toy, namely *speilzeug*, literally means "play-tool." A toy is no more than a specialized tool, designed and crafted for the purpose of playing some kind of games. The notion of *toys as tools* should not be limited to the like of toy-guns, and dolls; but should be applied broadly to cover other props and accessories, including dress-up costumes, game boards, and rolling dice.

More importantly, many game designers already knew that game *playing* is not determined by the toy, but by the players. It is the gamers who decide how the toy will be used in a game, and they often invent new ways to use the tools. Game modding being the case in point, gamers could either mod new contents – as in *Neverwinter Nights* (2002), *Halo: Combat Evolved* (2001); or uncover hidden content – as in the case of the "Hot Coffee" mod in *Grand Theft Auto: San Andreas* (2004).

1.4 Otium vs. Negotium

The last step in this chapter goes one step further, going back to the worldview of the forefathers of the Western civilization again following the path of words. Ancient Latin and Greek cultures organized all human activities into two categories. The free citizens of Rome and of the Greek city-state (meaning not slaves and foreigners) deserved to spend their days in what we would call "free time", i.e., time that could be spent in self-selected activities for the purpose of spiritual, intellectual or physical personal development. Such activities were termed *otium*, and included arts, sports, study, flirting, and children's games.

From the perspective of this chapter, *otium* mainly included autotelic activities, while all remaining activities were performed out of economic necessity, including fieldwork, craftsmanship, commerce, etc., and were classified as *negotium* (literally: the *neg*-ative form of *otium*). The dignity of free men would be diminished if they were compelled to spend their time for low or animal needs such as feeding and making money. Nevertheless, *negotium* was necessary for living, and was therefore assigned to slaves[6].

Already at a first look, games and playing, just like music and sports, were included in the *otium* part of life, and were therefore considered activities for the free man. Also, education, even if in forms quite far from the modern school system, was of course the peak of the activities aimed at personal development an grouped under *otium*. Education was the way to raise new generations of free man that would bring new life to the city. Indeed, Latin marks a striking relationship between *playing* and *education* as forms of *otium*.

---

[6] The value and dignity of the daily work as professional activity and the consequent affirmation of the dignity of slaved beyond the distinction of *otium* and *negotium* was first introduced by Christianism.

1.4.1 School and Teachers

In the last section we mentioned the Latin word *ludus*, which was used for institutional or performance games. A most interesting finding, perhaps even shocking to some readers, is that the same word belonged to the world of education as well. Teachers were in fact called *magister ludi* (literally, *Game Master*), and what we could call today schooling was called *ludus* (literally, Game).

Obviously, the notion of school in Ancient Rome is markedly different from today's public state schools (Gutek, 1995). In that world, education was the means to groom children into real men and citizens of the nation, through the tutelage of a wise teacher, as in the case of Alexander the Great, who was tutored by Aristotle. In this sense, education or schooling (*ludus*) was an institutionalized mean to engage youngsters in autotelic activities meant for the development of a free person. While historical differences are great and would require hundreds of pages to be accounted for, the ideal aim assigned to the education system is strikingly close to our perception of what schools should be made for.

Ancient Greek had an even stronger connection between *games* and *school*. The Greek equivalent for *otium* was *skolé*, which is actually the root word for *school* (Estienne, 1825). The Greek word *paideia*, which meant *game*, was also used to mean *education*, or the "upbringing of children". The word *paideia* is still to be found in the current English word *encyclopedia*, (*encyclo-paidedia*) meaning "all-round(ed) education." Interestingly, *paideia* came from the root words, **pai*, as in the Greek word *paizó*, which mean "playing" or "making a funny trick" (just like the Latin word *iocus*).

The modern idea and institutional practice of schools and education have gone through great semantic transformations and have departed from the worldview of ancient Romans and Greeks. Yet, the Greek and Latin cultures remain the roots of the Western civilization and of its idea of person, knowledge, society and education. At that beginning of our history, the words used for describing education, games, playing, upbringing of children, and learning are all very closely related. At that time and place, *school* was the means and an opportunity for young men to voluntarily submit themselves to a set of rules as a price to become "cultivated free individuals to who were suited to live in a free city" (Gutek, 1995, p. 28). While the activities thus became the end – an exercise to grow up as free men. This leads to the ninth implication.

> **IMPLICATION #9: School as Social Responsibility**
>
> The understanding of *going to school* in ancient Rome or Athens is far from being a superimposed mandatory participation to someone else's plans, but carries with it a social responsibility. Kids who attend school will learn the higher way of the society through *otium* and will join the society as free men, i.e., free from *negotium*, and therefore not slaves. Such children were held responsible for not wasting the opportunity: they were charged with the moral duty to perform (*play*) well in their learning. This probably echoes what we would like schools to be – an opportunity for raising free men and women – but is often at odds with the reality of many situations, in which school equals mandatory and boring. Also, school is often viewed as work (*negotium*): e.g. school work, homework, "Work on the Mathematics problems," etc., and not as a chance of expressing oneself.

1.5 Conclusion

Defining the word "games" is a difficult task (see Salen & Zimmerman, 2004). Hence, instead of trying to provide a definition for the term in the usual way (in a sentence or a paragraph), we have tried to examine several aspects about "games" and its relationship to "play," to gain a better understanding of the characteristics of games and how these characteristics affect education.

Firstly, we tried to refine our generic idea of *game* and *playing* into a more useful working concept for educators, bringing together several critical points about the use of games in the classroom. This was strengthened by a literature review about games as instructional devices or strategy that reviews the educators' viewpoints of "what games are," culled from textbooks and reference books for teacher education, technology integration into classrooms, instructional design and educational technology.

Secondly, we explored some modern disciplines that include the word *game* as a standard and well-defined concept in their field. Among the many possible, we selected Economics, Argumentation Theory, Semiotics and Philosophy. Each discipline defines *game* in a different way, highlighting specific features or aspects that contribute to a better understanding of the whole phenomenon. The first two steps paved the way for a more in-depth exploration of the complex meaning of playing and games in the Western tradition through the analysis of languages.

Thirdly, we compared the words used for these concepts in a sample of modern Western languages, namely English, French, Italian, German and Spanish. Finally, we provided an analysis of the meaning of the words: "games" and "play," by examining their meaning in the ancient Greek and Latin root words. Our analysis unveiled a most astonishing connection between "playing" and "education" as described by the Western classical worldview.

Based on the (nine) implications drawn during the chapter, it can be said that games are free-form activities that exist within a highly structured environment, to be enjoyed freely at certain moments in life. To do so, players are expected to voluntarily enter the game world and commit themselves by following the rules set out towards its end in order to resolve the tensions into a heightened sense of being. Toys are tools to assist in the playing, and though useful, they cannot dictate how the game is to be played because only the players can decide the course and the end of a game.

Even though our path of analysis has yielded many implications for games and play, we feel that the work is far from complete. As videogames research becomes increasingly acceptable (Hill, 2005), even more disciplines will join in the foray to *define* and *redefine* the concepts of *games* and *play*. The lexical analysis research frameworks provided in this chapter can always benefit from a more thorough treatise. Other researchers may want to affirm this work by extending the lexical analysis to include other ancient or modern Western languages, and even non-Western ones (such as Chinese, Japanese, Indian, African dialects, Slavic languages, etc.).

1.5.1 Education as Game

The notion of teachers as *magister ludi* (literally, Game Master) may also be worthy of further investigation. Loh (2007) suggested that teacher could lead an online role-playing videogame played by his or her class in the role of a Dungeon Master (DM; otherwise known as Game Master)[7]. This perspective seems to come close to that exploring the idea of aesthetic experience as paradigm for effective learning proposed by Parrish (2006), which also pointed at narrative strategies for instructional design. Even though the toolkit for Dungeon Mastering[8] has been available for some time, Loh (2007) noted that there has yet to be any work (or research) using this tool to support learning with videogames.

Huizinga wrote, "Let my playing be my learning, and my learning be my playing." The insights from our inquiry revealed that ancient societies view school, games, play, and education to be much more closely related that we would think. This is probably a first key point in the research on game-based education: many

---

[7] Dungeon Master is a term originated from the fantasy role-playing games, Dungeons & Dragons (King & Borland, 2003).

[8] To our knowledge, only the role-playing game called *Neverwinter Nights* (2002) offered such a DM Toolkit. Retrieved January 4, 2007 from http://www.gamespot.com/pc/rpg/neverwinternights/review.html

of the oppositions – between learning and playing, between fun and school performance – are possibly only optical illusions due to our distorted perspective.

In retrospect, we (the authors) have just reinvented the wheel – but this was probably a wheel that lay unused for such a long time as to be almost forgotten. By rediscovering the old meanings of the words within our tradition, we hope to help seeing the new context of game-based education and videogames in a clearer light. We hope we have provided insights and possibilities for educators, researchers, and game designers to scout new and innovative ways to better "*ludere*" our young people.

1.6 Acknowledgments

Luca Botturi wishes to thank Chiara Piccini for sharing her work about games and cultures. Special thanks go to Giuseppe Botturi for his help in consulting lexicographical works for ancient Greek and Latin.

1.7 References

Alessi, S. M., & Trollip, S. R. (2001). *Multimedia for Learning: Methods and Development* (3rd ed.). Needham Heights, MA: Allyn & Bacon.

Anglin, G. J. (Ed.). (1995). *Instructional technology: past, present, and future* (2nd ed.). Englewood, CO: Libraries Unlimited.

Aristotle (about 350 B.C.). Nicomachean Ethics. *Journal.* Retrieved from http://classics.mit.edu/Aristotle/nicomachaen.7.vii.html

Beck, J. C., & Wade, M. (2004). *Got game: How the gamer generation is reshaping business forever*. Boston, MA: Harvard Business School Press.

Bittanti, M. (2004). Per una cultura del videogames. Teorie e prassi del videogiocare. (Theories and praxis of videogaming for a culture of videogames). *Edizioni Unicopli, 2004*.

Callois, R. (2001, 1981). *Man, Play, and Games*. Champaign: University of Illinois Press.

Fincher, D. (Writer) (1997). The Game. USA: Polygram Films.

Gadamer, H. G. (1965). *Wahrheit und Methode*. Tübingen, Germany: JCB Mohr.

Gee, J. P. (2003). *What video games have to teach us about learning and literacy* (2nd ed.). New York: Palgrave Macmillan.

Grand Theft Auto: San Andreas. (2004). [Playstation 2 Games]. Edinburgh, U.K.: Rockstar North.

Gredler, M. E. (1996). Educational games and simulations: A technology in search of a (research) paradigm. In D. H. Jonassen (Ed.), *Handbook of Research for Educational Communications and Technology* (1st ed., pp. 521-539). New York: MacMillan.

Gredler, M. E. (2004). Games and simulations and their relationships to learning. In D. H. Jonassen (Ed.), *Handbook of Research on Educational Communications and Technology* (2nd ed., pp. 571-581). Mahwah, NJ: Lawrence Erlbaum.

Gutek, G. L. (1995). *A History of the Western Educational Experience* (2nd ed.). Prospect Heights, IL: Waveland Press.

Halo: Combat Evolved. (2001). [XBox Games]. Redmond, WA: Microsoft Game Studios.
Heinich, R., Molenda, M., Russell, J., & Smaldino, S. (1999). *Instructional Media and Technologies for Learning* (6th ed.). Columbus, OH: Prentice-Hall.
Heliö, S. (2004, February 19-22). *Role-Playing: A Narrative Experience and a Mindset.* Paper presented at the Solmukohta 2004, Helsinki, Finland.
Hill, M. (2005, September 25). More colleges offering video game courses. Retrieved September 26, 2005, from http://www.usatoday.com/tech/products/games/2005-09-25-video-game-colleges_x.htm
Howard, R. (Writer), Hallowell, T., Kehela-Sherwood, K., Grazer, B. & Howard, R. (Producer) (2001). *A Beautiful Mind.* [motion picture]. USA: Universal Picture Distribution.
Huizinga, J. (1980). *Homo ludens: A study of the play-element in culture* (R. F. C. Hull, Trans.). London: Routledge & Keegan Paul.
Johnston, J. (Writer) (1995). Jumanji. USA: TriStar Pictures.
Jonassen, D. H. (Ed.). (1996). *Handbook of Research for Educational Communications and Technology* (1st ed.). New York: MacMillan.
Jonassen, D. H. (Ed.). (2004). *Handbook of Research on Educational Communications and Technology* (2nd ed.). Mahwah, NJ: Lawrence Erlbaum.
Kahn, C. H. (1981). *Art and Thought of Heraclitus: An Edition of the Fragments With Translation Commentary.* Cambridge, UK: Cambridge University Press.
King, B., & Borland, J. (2003). *Dungeons and Dreamers: The Rise of Computer Game Culture from Geek to Chic.* Emeryville, CA: McGraw-Hill/Osborne.
Koster, R. (2005). *A theory of fun for game design.* Scottsdale, AZ: Paraglyph Press.
Lever-Duffy, J., McDonald, J. B., & Mizell, A. P. (2003). *Teaching & Learning with Technology.* Boston, MA: Allyn & Bacon.
Loh, C. S. (2007). Designing Online Games Assessment as "Information Trails". In D. Gibson, C. Aldrich & M. Prensky (Eds.), *Games and Simulation in Online Learning: Research and Development Frameworks* (pp. 323-348). Hershey, PA: Idea Group, Inc.
Maddux, C. D., Johnson, D. L., & Willis, J. W. (1992). *Educational computing: Learning with tomorrow's technologies* Boston, MA: Allyn and Bacon.
Mann, W. C. (2002, July 11-12). *Dialogue Macrogame Theory.* Paper presented at the 3rd SIGdial Workshop on Discourse and Dialogue Philadelphia, PA.
Melchiorre, V. (2006). Gioco. In V. Melchiorre (Ed.), *Enciclopedia Filosofica.* Milano, Italy: Bompiani.
Michael, D., & Chen, S. (2006). *Serious games: Games that educate, train, and inform.* Boston, MA: Thomson Course technology PTR.
Nash, J. F. (1951). Non-Cooperative Games. *The Annals of Mathematics, 2nd Ser., 54*(2), 286-295.
Neverwinter Nights (Version 1). (2002). [Computer Games]. New York, NY: Atari, Inc.
Parrish, P. (2006). *Aesthetic Principles for Instructional Design.* Paper presented at the annual conference of the Association for Educational Communications and Technology (AECT 2006).
Prensky, M. (2000). *Digital game-based learning.* New York: McGraw-Hill.
Rieber, L. P. (1996). Seriously considering play: Designing interactive learning environments based on the blending of microworlds, simulations, and games. *Educational Technology, Research, and Development, 44*(2), 43-58.
Rieber, L. P., Smith, L., & Noah, D. (1998). The value of serious play. *Educational Technology, 38*(6), 29-37.
Rosen, L. D., & Weil, M. M. (1995). Computer Availability, Computer Experience and Technophobia among Public School Teachers. *Computers in Human Behavior, 11*(1), 9-31.
Salen, K., & Zimmerman, E. (2003). *Rules of play : Game design fundamentals*: The MIT Press.
Salen, K., & Zimmerman, E. (2004). *Rules of play: Game design fundamentals.* Cambridge, MA: The MIT Press.

Schiller, F. (2004). *On the Aesthetic Education of Man* (R. Snell, Trans.). New York: Courier Dover Publications.

Sheldon, L. (2004). *Character Development and Storytelling for Games*. Boston, MA: Thomson Course Technology PTR.

Smith, P. L., & Ragan, T. J. (1999). *Instructional design* (2nd ed.). New York: John Wiley & Sons, Inc.

Smuts, A. (2006). Video Games and the Philosophy of Art. *Journal*. Retrieved from http://www.aesthetics-online.org/ideas/smuts.html

Van Ments, M. (1999). *The effective use of role-play: Practical techniques for improving learning* (2nd ed.). London, UK: Kogan Page.

Von Neumann, J., & Morgenstern, O. (1953). *Theory of games and economic behavior*. Princeton, NJ: Princeton University.

Walton, D. N. (1984). *Logical dialogue-games and fallacies*. Lanham, MD: University Press of America.

Chapter 2
Learning by Designing Homemade PowerPoint Games

Lloyd Rieber, Michael Barbour, Gretchen Thomas, and Dawn Rauscher

Abstract

There are two fundamental ways to approach the use of gaming in education: playing games or designing games. In this chapter, we take the second path and discuss how to use an almost ubiquitous software tool in the schools — PowerPoint — with children in the design of original games. In these games, students put school's content to use in ways perceived to be meaningful and authentic to them. Game design capitalizes on the natural design instincts of children. Children are natural game designers who tacitly know at deep levels what makes a good game and what makes a game fun. Turning over to them the challenge of designing a game that also teaches something is a worthy design goal and is aligned with the goals of schools and teachers. We discuss the balance between learning and motivation that must occur in the act of design and use play theory and flow theory to help explain this. Good games provide among the best examples of cognitive psychology in action: authentic goals situated in a meaningful and motivating activity with clear and consistent feedback. However, ideas and approaches, no matter how innovative, which cannot meet the practical demands and limitations of a typical school are doomed to failure. Consequently, our approach meets head-on the issue of scalability, that is, having the idea take hold and flourish within the limited resources of a typical school with teachers who already have too many expectations placed on them.

2.1 Introduction

A typical school day for most children is filled with intellectual, emotional, and social challenges, too much of which is negative. Intellectually, children feel the tension arising from the current emphasis on high-stakes testing. They know that much is riding on their test performance. Some do well and some do poorly, but most are just happy to have survived yet another round of testing and judgment by adults. Those who do well in school must be careful that news of their success does not reach their peers because succeeding in school is often the nemesis of popularity. Many students don't care about their school performance, despite the threats or punishments. Perhaps it is naïve or idealistic to think that school should be a place of wonder and excitement, but we hold such a view. We envision school as place where learning resides at the intersection of intellectual curiosity, emotional well-being, and feelings of social worth. Such a place is likewise full of challenges, but ones that students meet with excitement and commitment. Our philosophy, and our strategy, to reach these lofty goals are based on design activities, that is, learning as a result of the act of designing and building something.

Of all the possible design artifacts that could be built, we focus exclusively on games. Designing a game requires much more than just the accumulation of accurate content, as important as that might be. A good game is designed based on relationships of the content over time, much in the same way as a good story takes time to unfold. A good game is also engaging, and, well, fun to play. This suggests a quality standard that is hugely influenced by one's peers. No one would want to design a game that one's classmates consider boring. This inherent tension of seeking "good design" as defined by, or recognized by, one's peers carries significant advantages when designing a game that also has explicit educational goals. The challenge of finding ways to use and apply the content in compelling ways becomes a design challenge for students, not the teacher. Of course, a student must understand the content in order to do this — that's point, and the advantage, of a design approach.

It is also crucial for teachers to feel invested and capable as the design process unfolds. The challenges that teachers confront each day are huge and easily downplayed by non-teachers. Getting permission to use software beyond what administrators and technology coordinators have approved for installation on classroom or lab computers is a battle that most teachers are just not willing to fight. Of all the software tools available for education today, one tool that is likely to be available in almost every school in America is Microsoft PowerPoint. Although there are certainly much better game design tools, we believe that harnessing PowerPoint for game design is a practical way to bring game design into the classroom. We refer to the design artifacts resulting from this approach as "homemade PowerPoint games." The words "homemade" and "games" capture well the personal nature of the design task. The word "PowerPoint" obviously refers to the tool, and we hope it is clear that this word can easily be replaced with other tools that a teacher may readily have now or in the future.

In this chapter, we present our vision, theory, philosophy, and evolving strategy for creating and playing homemade PowerPoint games in the classroom.

2.2 Game Design: The Other Use of Gaming in Education

The word "game" is a loaded term, especially in educational settings. One rarely finds an educator, administrator, parent, or legislator with a neutral opinion on the role of games in education. Many believe that games are fine, but only as a reward for completing one's work. Turning something into a game can also denote a sense of degrading or belittling of the content. Many people equate gaming and technology with commercially available video games, a media form often associated with violent content. No one wants such material to enter the school building. It is true that many of the commercially available games are tasteless and violent, but it is important to remember that the term "game" refers to a media genre or art form, just as do the terms "book," "movie," or "cinema." And, just like books and movies, games can be inspiring, good, bad, or disgusting. In this project, it is important to remember that the focus is on designing games much more than playing games. We believe the use of gaming in this way is a serious, sophisticated, complex design activity that requires students to understand the richness and interconnectedness of the subject matter.

The motivational characteristics of games get a lot of attention, even by those who see themselves as game advocates (Dempsey, Lucassen, Gilley, & Rasmussen, 1993-1994). Games, almost by their very definition, are motivational because they are what people associate with leisure activities. The motivational characteristics of games are many: competition, random features giving rise to the unexpected, challenge, and curiosity. In four studies conducted by Ryan, Rigby and Przybylski (2006), they found motivation in gamers was largely based on the level of autonomy the game allowed, the challenge presented by the game, and the presence felt by gamers within the gaming environment. Good games give a player a feeling of control, but with an edge of uncertainty (Prensky, 2001).

However, the motivational side of gaming plays a small role in the design of homemade PowerPoint games. Instead, this approach focuses on the cognitive and cultural characteristics of games (Squire, 2002). First, games provide a meaningful, relevant context. The importance of context and situation has been underscored in education over the past 20 years. Some argue that learning must be situated in a meaningful context for learning of any value to occur (Lave, 1988). Games also organize content in story-like ways. Good games help players to see the relationships between all of the game objects. Anyone who understands the game Monopoly knows of the relationship between passing Go, buying property, building hotels, and going to jail. These are not disjointed events, but synchronized ones. They are affected both by chance, but also by decisions made by the players. There are good strategies and bad strategies. Good games also have clear goals, such as the object of the game, or what one has to do to win. Related to goals is feedback. According to Gee (2003), one of the reasons why video games are so motivating and engaging is because of their ability to provide "explicit information both on-demand and just-in-time, when the learner needs it or just at the point where the information can best be understood and used in practice" (p. 138). The best games make it clear throughout the game play whether or not the player is inching closer to or farther from success.

The role of organization, situation, goals, and feedback in a game is closely related to strong narrative structures in stories. Good games, like good stories, develop and balance these structures in ways that both provoke and tease. In both games and stories, we wonder what will happen next. Unlike a story, a game invites true participation in that the ending of the game will depend in large part on what we, the players, do. As Wolf (2001) describes, games "involve the audience in a uniquely direct manner, making the *viewer* into a participant or *player*, by allowing the player to control (to some degree) a character in the game's diegetic world" (emphasis in the original, p. 93). Grodal (2003) goes even farther, stating that the story in the game "is only developed by the players active participation" (p. 139). Both of these views are consistent with constructionism, where the individual (in this case the game) is involved in the art of building the narrative.

Using games in education as a route to learning can be conceptualized two ways: playing educational games designed by others or designing your own game. The research literature focusing on whether playing games leads to learning (gains in achievement) is mixed (Kirriemuir & McFarlane, 2004). When games are compared to traditional classroom instruction (the most common research method), few differences in learning are reported (Dempsey, Lucassen, Gilley, & Rasmussen, 1993-1994; Gredler, 2003; Randel, Morris, Wetzel, & Whitehill, 1992). However, this is consistent with the "no significant differences" problem that usually occurs when innovative educational technologies are compared with traditional approaches and has been well documented by Clark (1983), Reeves (2005) and Russell (1997) among others.

Another approach to gaming focuses on what students learn from designing their own games. As Prensky (2006) states, games such as *Age of Empires* and *The Sims* have built in features that allow for players to modify, or "mod," the game's characters, environment and even the game play. Essentially, allowing game players to become game designers. Gee (2005) describes modding as ranging "from building new skate parks in *Tony Hawk* or new scenarios in *Age of Mythology* to building whole new games" (p. 35).

A good example of this research on gaming comes from the work of Yasmin Kafai (1994, 1995) and her colleagues (Kafai & Harel, 1991). Their research has focused on student motivation and learning while building multimedia projects. In these "children as designers" studies, elementary school students are typically given the task of designing an educational game for a younger audience (i.e. fifth graders designing for third graders). In one example (Kafai, Ching, & Marshall, 1997), qualitative results showed how students used the design activity as an opportunity to engage in content-related discussions. Quantitative results also demonstrated increased learning of astronomy concepts by students.

For over a decade, we have spent time with elementary and middle school students designing and developing educational computer games (Rieber, Luke, &

Smith, 1998). The motivation for starting the project was based on a simple question: Would children be able to take advantage of the opportunity to engage in game design when properly supported to do so? The purpose of our research was to observe the social dynamics of children working collaboratively in teams. Besides the qualitative data produced from our observations, our other primary measure was whether or not the children produced a working game. The games themselves became an artifact for research by giving us insights into how children find ways to put subject matter they were learning in school to use in a context that they value (i.e., games). This research showed that students were able to work effectively in groups to produce games that creatively embedded subject matter.

2.3 Constructionism as a Rationale for Gaming in Education

The intellectual justification for homemade PowerPoint games comes from the epistemology of constructivism. The term constructivist is used metaphorically (and often too loosely) to refer to learning as a process where individuals construct their own knowledge through meaningful interactions with the world. Learning is considered to be an active, controllable process that builds upon a student's prior knowledge and is grounded in meaningful, social contexts (Hooper & Rieber, 1995). This view is contrasted with transmission models of education where learning is viewed as the passing of knowledge from one person (e.g., teacher) to another (e.g., student) (Grabinger, 1996). Modern interpretations of constructivism have been influenced by the work of Piaget, Vygotsky, and Dewey (see Duffy & Cunningham, 1996, for a review). In this project, John Dewey's educational philosophy, especially his progressive views of democratic forms of education (Dewey, 1916), have been particularly influential. Students should have a say in what they learn and how they learn, what Papert (1993) referred to as the "right to intellectual self-determination" (p. 5). This does not mean that education should be a free-for-all without purpose, goals, or expectations, but it does mean that students should have a say in what transpires in the classroom. Specifically, in a democratic learning environment, students do not decide for themselves what will be learned, but rather have varying degrees of choices in negotiation with the teacher.

There are many forms of constructivist learning. In this project, we use a specific form called constructionism. Constructionism contains both a constructivist philosophy of education, but also a set of learning strategies based on its name – the act of learning by building. The relationship between constructivism and constructionism is best summed up by Seymour Papert (1991):

> "Constructionism — the N word as opposed to the V word — shares constructivism's connotation of learning as 'building knowledge structures' irrespective of the circumstances of the learning. It then adds the idea that this happens especially felicitously in a context where the learner is engaged in constructing a public entity, whether it's a sand castle on the beach or a theory of the universe". (p. 1)

Negotiation is also a common element in a design activity. When designing, one tries to build an artifact conjured up by the mind, but not all designs are possible given the tools, resources, support, and time available. Interestingly, design is really an exercise in exclusion – of all of the possible designs imagined, the task is to slowly settle on only one design to the exclusion of all others. Such an approach does not negate the need for instruction, but rather uses instruction in only the most deliberate of circumstances.

Over the past decade, a growing body of literature has supported the contention that children can learn in powerful ways when they are involved in the act of building projects (Blumenfeld et al., 1991) (Harel & Papert, 1991; Kafai & Resnick, 1996). So far, most examples of constructionism have been in mathematics and science. Through these approaches, students as young as middle school have grasped ideas from calculus (Roschelle, Kaput, & Stroup, 2000), geometry (Olive, 1998), and genetics (Horwitz & Christie, 2000).

As exciting as these projects might be, one of the main hurdles for any innovative project to overcome is the problem of scaling – that is, widespread implementation of the project in schools without special assistance or resources. The story of the problem of scaling told by researchers of innovative technological approaches is regretfully too common (Spiro et al., 2002). A team of researchers arrives at a site and is greeted enthusiastically by the teachers and administrators. They do wonderful things for weeks or months at a time. But, after the researchers leave, the school reverts to teaching as they did before the researchers arrived. Innovations not adopted by the teachers and supported by the administration are abandoned. Of course, the focus of these research projects was not on adoption and diffusion, but on learning and cognition. However, theory and research without practice does not make a difference in the long run because the innovation is not sustainable.

One technological innovation that has surmounted the scalability problem is the WebQuest, initially conceived by Bernie Dodge of San Diego State University (http://webquest.sdsu.edu/). A WebQuest project is a creative instructional use of the Internet that has been embraced by teachers. One reason for the success of WebQuests is that they are based on what teachers already do – design instruction for students – while making good use of Internet resources and making good use of student time. Teachers who design WebQuests feel satisfaction in integrating Internet-based resources in ways that are consistent with their training and daily school responsibilities.

There is no denying the success and popularity of WebQuests among teachers (Lamb & Teclehaimnanot, 2005). For those of us interested in technology integration in the schools, this is a significant step in the right direction. But the true genius of Bernie Dodge in developing the WebQuest idea is not with the activity itself, but with the degree to which teachers have adopted the technology. WebQuests have been successful because, like a beneficial virus, they have spread among teachers. It's hard to say why this is exactly. One reason is that designing a WebQuest takes teachers but a short step away from the comfort of their ordinary teaching routines. It challenges them just enough to feel satisfying and it uses technology in ways that they can understand and use. The adoption of WebQuests by such a large number of teachers and schools is a significant milestone in technology integration, if only because the decision to accommodate the innovation was made by the teachers themselves. This is an example of a technological innovation overcoming the challenge of scalability — widespread adoption without significant increases in resources. However, if one looks closely at what a WebQuest actually entails, one finds that it is by and large an instructivist example of technology integration – most WebQuests are web-enhanced forms of direct instruction. Those of us interested in constructivist forms of learning are not satisfied with WebQuests and yearn for something much more student-centered and student-directed.

2.4 Game Design as a Route to Play and Flow

We see game design as a valuable activity because it aligns what school values to what children value. We believe that schools and teachers value much more than merely an ever-increasing score on a standardized test, despite the political pressures for their students to do well. We believe that schools and teachers want children to be engaged in the curriculum, excited by it, and to see the wonder and elegance of all the content areas. Pre-school children grow up seeing the world as "wonder-full" where opportunities abound to continually trigger then satisfy their curiosity and need to learn. Unfortunately, the life of a child in and out of school is usually pitted against each other. Design activities provide children with control, negotiated with the teacher, over finding uses for content and for representing that content and relationships within it in meaningful ways. We have used play theory and flow theory to guide our thinking about children as game designers, the combination of which we have termed "serious play" (Rieber & Matzko, 2001; Rieber, Smith, & Noah, 1998).

It is important to distinguish the cognitive and emotional definition of play from a purely behavioral one. That is, a teacher can take students to the gym and have them "play" a game, but they may not be experiencing the play phenomenon. Play, in the sense that it is used here, is a state of mind, not one solely of behavior. Children who are truly "at play" are completely absorbed by the activity. Their attention is devoted to it with the sense of being freed from the non-play world. It's as though play is a special space in which things can be done which one could not do otherwise – after all, it is only play. Play theory can be organized around four themes, or rhetorics (Sutton-Smith, 1997): Play as progress; play as power; play as fantasy; and play as self. Designing games for learning can involve all four. Play as progress is the belief that play leads to something else, usually something beneficial. That is, there is a belief that play leads to a tangible outcome. For example, a child who plays "grocery store" is believed to be learning about money and food choices. This is consistent with the Piagetian view of play serving the role of assimilation and imitation serving the role of accommodation. Game design as play leads children to put school knowledge to use. Play as power involves competitions or contests in which winners and losers are declared. Though competitive behavior can, of course, can have negative consequences, competition is central to gaming. Designing a game where players enjoy the competition without feeling humiliated can be a healthy design challenge. Play as fantasy involves make-believe scenarios, contexts, and people. This is the creative side to play and all good games benefit from creative design. Play as self may be the most important in the sense that here play is considered an optimal life experience which triggers the flow phenomenon (Csikszentmihalyi, 1990, 1996). The act of play is its own reward and outcome. This is obviously most aligned with theories of intrinsic motivation (Deci & Ryan, 1985; Lepper, Keavney, & Drake, 1996). Children who experience play as self engage in game design because they see it as intrinsically worth their while.

Flow theory is defined by Csikszentmihalyi as an optimal life experience in which people are so absorbed and engaged by the experience or activity that time seems to vanish — they are carried along by the flow of the activity. The similarities between play and flow are striking. Like play, the flow experience is elusive but can be triggered at almost any moment. Adults typically report experiencing flow in their leisure-related activities, such as gardening or woodworking, but any activity can produce it. Flow theory includes the following components:

- The challenge of the activity is optimal;
- The person's attention is completely absorbed by the activity;
- The activity has clear goals and provides clear feedback on whether the person is meeting those goals;
- The activity is so absorbing that the person feels free of other worries;
- There are no feelings of self-consciousness;
- The person feels in total control of the activity;
- Time does not seem to exist.

Like play, flow requires total attention on the part of the individual. In fact, this point is so important that Csikszentmihalyi considered attention to be the "fuel" for the flow experience itself. In a design activity, attention is focused by the use of a "driving question" (Blumenfeld, et al 1991; Grant, 2002) that leads the design, such as "why are the necks of giraffes so long?" or "why is it cold in January in the northern hemisphere, but warm in the southern?" Children understand the "come up with a good game" task at an intuitive, deep level and this becomes their driving question when designing a game.

2.4.1 Homemade PowerPoint Games: How Do They Work?

We use PowerPoint as a game design tool. Although there are better game design tools, such as StageCast, PowerPoint has three distinct advantages: it is already available in most classrooms; its use by teachers in any form is directly associated with technology integration; and teachers like using it. PowerPoint is an almost ubiquitous software application in schools. Teachers do not need to ask anyone's permission to use it and once installed it is a very robust and reliable application. This "transparency of use" is vitally important and often overlooked by educational technologists. For example, we tried for many years to encourage the teachers who went through our graduate program at the University of Georgia to learn and use StageCast (http://www.stagecast.com/), a very creative and innovative software application specifically designed to allow students to build interactive games and simulations. Teachers liked learning it, but they clearly did not see it fitting into their classroom routines. There were two reasons for this. First, to install StageCast at their schools meant that it would have to be purchased and installed. StageCast is very inexpensive, but it is not free — teachers would either have to buy it themselves or have to convince their administration to purchase it. Beyond the expense was the hurdle of getting it installed on computers in their schools. Though seemingly a small hurdle, teachers express much frustration in the amount of time it takes to work with system administrators and technology committees to allow isolated software applications to be installed on a school's computer.

We view PowerPoint as an "intellectual transitional" software application. By this we mean that its use is accepted by the current culture of school, but can also act as a bridge to uses outside of that culture. It is accepted because it is intended primarily as a presentation tool and, as we all know, teachers are expected to make presentations. Teachers are therefore rewarded by using it even by those administrators who take a very traditional, transmission model of education. But as a teacher masters PowerPoint as a presentation tool, it becomes relatively easy to coax them to use it in other ways. Interestingly, one of its most popular other uses is as a gaming tool, though not one we support. This gaming use is largely for the coverage of decontextualized factual knowledge using popular television game formats or templates, such as those based on the game Jeopardy. These templates have been widely disseminated among teachers. Again, we are not very excited by this gaming approach, except that it takes one step away from the presenting of information to the interaction with information. Being a small step, both educationally and technically, teachers are very comfortable with this use, so we build on this interest of those teachers who use PowerPoint in this way. Figure 2.1 shows the progression from an instructionist to a constructionist perspective in using PowerPoint games in the classroom. It is also important to note that the only new and challenging technical skill teachers need to master to design games with

PowerPoint is how to make hyperlinks. However, teachers and students easily learn this skill and all find it a very satisfying accomplishment.

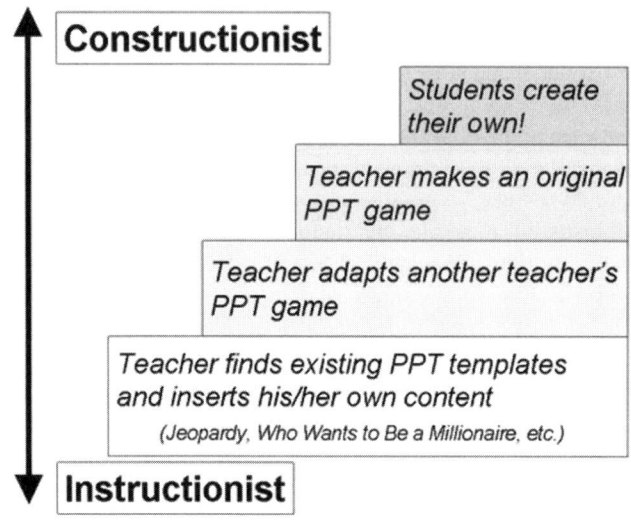

Fig. 2.1 The progression from an instructionist to a constructionist perspective in using PowerPoint games in the classroom. The goal of this project is for teachers to reach the top of the hierarchy in their classrooms.

2.4.2 What does a homemade PowerPoint game look like?

Like most games, homemade PowerPoint games have the following elements: game world, story or context; game play or goals and strategies within the game, and game rules (Aarseth, 2003). Probably the best way to understand what a homemade PowerPoint game looks like is to consider an example. Lloyd Rieber designed and built a game called "The Traveling Georgia Artist." All homemade PowerPoint games start with writing a story. A good story not only hints at how to play the game, it also situates the game in an interesting and understandable context. Here is an excerpt of the story of the traveling Georgia artist, featuring Chris, a poor teenager who lives in rural Georgia who dreams of attending college and becoming a great artist. To do so, Chris comes up with a strategy to earn enough money for college[1]:

---

[1] This game can be downloaded at the following URL: http://projects.coe.uga.edu/lrieber/wwild/search/module-detail2.asp?ID=753

Chris decides to borrow a friend's old van and travel all through the State of Georgia selling the works of art Chris has made. If Chris can earn enough money, Chris will attend the University of Georgia in the fall semester. But, if unsuccessful, Chris will have to return home to help out in the family business. This is Chris's big chance to see Georgia, earn a reputation as a real artist, and get the opportunity to go to College.

In order to play the game, the players have to print out and tape together four slides to form a map of the State of Georgia. This map contains the major cities of Georgia along with the major Interstate highways. There is also a grid system superimposed onto the map. The game begins by players opening a bank account containing $500 and starting their journey by placing their game piece anywhere on the map. During the game, the players move their game pieces on the map according to the number of moves indicated by a roll of the dice. Players can either move one square at a time, or they can move in large increments by traveling on the Interstate highways. After every turn, they have to select an "Expense" card that shows their expenses for the day. If they land on a city, they can select a "Sell Art" card that shows what artwork they sold that day and for how much. Also included on the game map's grid are spaces marked "quiz." If a player lands on one of these, attention is turned to the quiz section of the PowerPoint file. A player chooses one of the programmed game questions. If answered correctly, the player earns double their last sale of art works and the question is marked with an "X" on a printed copy of the slide to indicate that this question can no longer be selected. If answered incorrectly, the question remains in the pool for other players to try to answer. Obviously, other players benefit from seeing what the question is and also what is an example of one of the incorrect responses. All players keep a checkbook while playing the game. A player wins by either being the first to earn a total of $2000, or is the last person not to declare bankruptcy.

It is important to understand that a game can only be understood well by playing it. Also, a good game is designed only after refining it carefully over time while playing it and discovering what creates optimal game play. For example, in the case of the Traveling Georgia Artist, it took many revisions to discover what dollar amounts facilitated game play and game enjoyment on the respective expense and sell art cards. A good example of the importance of game play for revising the game is a rule that was designed into the game after Lloyd Rieber played with his daughter, Rebecca, a professional graphic designer who inspired the game's original idea. While playing the game, it happened that one player landed on top of another. This seemed to be a special moment in the game and it seemed that something ought to happen. Rebecca had the idea that the person who was landed upon should be required to take and pay for an art lesson, with the money going to the person who landed on top. This rule embeds a whole class of strategies and healthy tension into the game and improves it considerably.

2.4.3 Homemade PowerPoint Games are an Example of Appropriate Technology

By appropriate technology we mean that all everyday objects can be incorporated into the game. There are no rules that say that one part should or should not be played on the computer. The entire game could be played on the computer, the entire game could be played off of the computer, or, as in the case of the Traveling Georgia Artist, some is played on the computer and some is not. However, in all cases, PowerPoint acts as the means for delivering, disseminating, and sharing the game to others. All game directions and resources (or directions on how to make resources from everyday objects) are included in the PowerPoint file. When you download the PowerPoint file, you have a blueprint for creating the entire game, usually just by printing certain slides and using other slides while playing the game. This design model encourages students to consider any everyday resource that could enhance the game.

It's also important to point out that although we focus much attention on PowerPoint in this project, for all of the reasons outlined above, PowerPoint is merely the means towards an ends. We hope that once teachers begin using PowerPoint with their students in a constructionist way to design games, they will then more readily make the jump to other tools, even those that do require them to persist in getting permission to be installed on their school's computers and servers.

This project also includes an "open source philosophy" to the sharing and dissemination of the games. Students and teachers can freely adapt existing games and even submit their new versions to the web site so long as credit is given to the previous designers. This is an important part of the project because of our belief that game play triggers creative game design. That is, ideas about improving a game usually only surface while playing the game. We want all students to see themselves as game designers, even if they are playing an existing game. So, ideas that would improve the game can be the impetus to creating a new game, adapted or based on an existing game. Students would only be required to give credit to the original designers of the game (on a credits slide) before uploading their new game to the Web site. This approach allows for very practical and efficient use of class time for game design. Teachers do not have to guide students always through the entire game design process from the beginning.

2.5 Homemade PowerPoint Games in Use: K-12 Education

Although this book is primarily focused on theoretical arguments for gaming in education, we take a holistic approach in which theory, research, and practice must be balanced together. Consequently, we end this chapter with a model for implementation that is consistent with the research and theory. Our work suggests that the successful integration of homemade PowerPoint games in a K-12 classroom requires a 6-stage approach: 1) Student orientation; 2) student game design — scaffolding by the teacher; 3) student story writing; 4) developing an early prototype; 5) refining the prototype; and 6) iterative cycle. Each stage will be briefly described in this section.

2.5.1 Stage 1: Student Orientation

The teacher leads the class in a discussion of game design. The teacher starts by asking students if they play games, and if so, what games do they play. Some examples of discussion questions include: 1) What are examples of games that you used to play as a small child?; 2) Why do you no longer play those games?; 3) What makes a game fun?; and 4) What is the difference between a good game and a bad game? The purpose of these questions is to help students begin to articulate examples of game design elements, such as the need for optimal challenge.

The next step in the student orientation is for the class to play an existing homemade PowerPoint game. As shown in Figure 2.2, this is the first step to preparing students for eventually creating their own PowerPoint games. The teacher will decide which game is to be played. The purpose of this activity is to acquaint students with the form and style of a homemade PowerPoint game. After the class plays the game, the teacher will lead a discussion of how this kind of game is different from a video game, but that this type of game is something that they can create themselves. The teacher may decide to have the class play other existing homemade PowerPoint games during several classes until the teacher feels that the students have a firm grasp of the style and format.

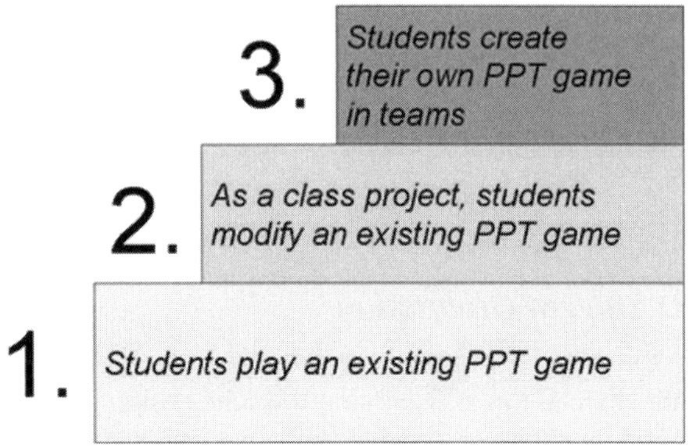

Fig. 2.2 The progression of phases that students need to complete in order to be prepared to create their own homemade PowerPoint games successfully.

1. Students play an existing PPT game
2. As a class project, students modify an existing PPT game
3. Students create their own PPT game in teams

2.5.2 Stage 2: Student Game Design — Scaffolding by the Teacher

In this stage, the teacher guides the students in recommending changes to one of the homemade PowerPoint games they played. Students may suggest several games, so the teacher needs to eventually focus attention on just one of the games. The goal is to evoke game design ideas from students that would improve the game and make it either more entertaining or more educational. The goal of this activity is to show students that they take an existing game and revise or adapt it in some way. The teacher explains that this is permissible so long as they give credit to the original designers (this is part of the "open source" philosophy of the project).

2.5.3 Stage 3: Student Game Design — Student Story Writing

In this stage, the teacher requires each student to write a story pertaining to the content studied so far in the class. The teacher will remind students that these stories will be the basis of their homemade PowerPoint games. The teacher will decide how long to give the students to write their stories, though it is recommended that the students take their time with the activity and also do much of their writing as homework.

After the students have written their stories, they read their stories to the class, either to the whole class or in small groups. The teacher then leads a discussion with the students to decide which stories might be the best for game development. The teacher should be careful to point out strengths of all the stories while also leading students to see which stories effectively embed historical relationships and lead to an interactive format.

2.5.4 Stage 4: Student Game Design — Developing an Early Prototype

In this stage, the teacher divides the class into small groups where each group acts as a design team and uses one of the stories written by the students as the basis for their game. Because this is an early attempt at game design, the teacher may purposely not want to impose much structure on the student groups, even though it is likely they will have difficulty in completing the game design phases effectively. The goal of this experience is to orient student to working in groups and to give them some idea of the time it takes to develop the game. It will also show them the need to get organized in order to complete the game design phases within the time given.

2.5.5 Stage 5: Student Game Design — Refining the Prototype

The teacher gives students a deadline for completing a working example of their game design. Those that produce a working example can showcase their game to the rest of the class followed by the entire class playing the game (usually in small groups).

Those students who did not produce a working example by the deadline will be a little frustrated. However, using this experience, the teacher will lead a design debriefing. The outcome of the debriefing will be lessons learned by the students. A good recommendation is to have team members take on specific roles in the next game design attempt. Typical roles include project manager, lead designer, lead developer, and content expert.

The teacher will also ask the class if they feel any of the games are of sufficient quality to be uploaded to the WWILD Team Web site (http://it.coe.uga.edu/wwild/) to be shared with other classes. It is unlikely that the games produced in this first attempt will be suitable for sharing, but this provides the teacher with a way to remind students that they can become published game designers if they learn from this experience.

2.5.6 Stage 6: Student Game Design — Iterative Cycle

Based on the experience gained by students up to this point, the teacher then begins to use game design as a means for examining the next unit in their history curriculum. Students should study the unit with an eye towards how to develop an effective story that will embed the facts, characters, and places of the historical event. The teacher will lead discussions, both with the entire class, and with each of the small groups, that focus more and more attention on the historical content and less and less attention on the game development process. The rationale here is that the students should be becoming acclimated to the demands of the game design task, so that most of the design attention will be spent on finding creative ways to exploit the historical relationships of the unit in ways that promote more interesting games.

To date, our most extensive experience with homemade PowerPoint games is in the pre-service instructional technology courses that we offer to undergraduates at the University of Georgia. The game design unit includes five sessions, each one class period in length and roughly corresponding to the implementation model above. As a game design skill, students are introduced to the concept of different levels of questioning. These levels, based on Bloom's Taxonomy (Bloom, Englehart, Furst, Hill & Krathwohl, 1956), are compared and contrasted with the Levels of Historical Understanding (Wineburg, 2001). Students work in pairs to improve their three sample questions so that the games will promote higher-order thinking skills and interpretation of historical events for the targeted audience rather than the basic recall of facts. As per the game design implementation model, the unit ends with the students sharing games with the class. The final activity is a class discussion where these pre-service teacher education students discuss the role of game design in K-12 teaching and learning: How do teacher-designed games promote critical thinking among students? How do student-designed games promote critical thinking among students? When/where is each appropriate in an educational setting? Table 2.1 lists a rubric for the evaluation of the games designed by these students.

Table 2.1 Homemade PowerPoint Game Rubric Used in a College Pre-Service Education Course

-	Excellent	Average	Unsatisfactory	Total points
Student Appeal	Students in the targeted grade levels will be interested in all of the aspects of this game. (13-15)	Students in the targeted grade levels will be interested in some of the aspects of this game. (8-12)	Students in the targeted grade levels will be interested in few or no aspects of this game. (0-7)	___/15

-	Excellent	Average	Unsatisfactory	Total points
Enhances Education	The game greatly enhances the students' understanding of specific QCC's. (18-20)	The game has the potential to enhance one or two specific QCC's. (14-17)	The game is not designed to specifically enhance any QCC's. (0-13)	___/30
Educational Value	The game is valuable enough to devote classroom time to playing the game. (18-20)	The game has moderate educational value in the classroom. (14-17)	The game has little or no educational value in the classroom. (0-13)	___/30
User-friendly	All of the materials, the directions, the story and the objectives are accessible via the PowerPoint presentation. Everything is clear and easy to understand. (18-20)	Some of the materials, the directions, the story and the objectives are accessible via the PowerPoint presentation. Some things are not clear. (14-17)	Few of the materials, the directions, the story and the objectives are accessible via the PowerPoint presentation. Significant explanation is needed. (0-13)	___/10
Thinking skills	The game involves significant strategic components that enhance thinking and decision-making skills. (13-15)	The game involves some strategic components that enhance thinking and decision-making skills. (8-12)	The game involves few or no strategic components that enhance thinking and decision-making skills. (0-7)	___/15
Comments:			Total ___/100	

QCC: Quality Core Curriculum

2.6 Conclusions

The advances in technology – 3D processing, networking, data management of hundreds or thousands of players, to name by a few – have led to a gaming culture in our society that has never been seen before. Given the current state-of-the-art gaming technology and the Moore's law rate of advance, the future is bright for anyone interested in gaming and technology. It may be surprising to many, therefore, of our advocacy of such an apparent low-end gaming technology such as

PowerPoint. We want to see gaming used productively as a means that enable the ends of school and the desires and talents of children. To realize this goal, we have taken a path founded on two critical assumptions that distinguish our work from many others represented in this volume. First, we take a strong constructivist stance on gaming that leads to the position of children as designers of games instead of merely players of games. Second, we approach the topic from the practical point of view that considers and accepts the state of technology in the schools in order to meet directly the problem of scalability: What technology do schools have and teachers use? As we have stated in this chapter, our views on educational gaming balance theory, research, and practice. Consequently, our motives and our ideas are informed by practice with a constructivist ambition. So, our seemingly low-tech approach is contrasted by our audacious ambition to change activities in schools from teacher-centered to child-centered based on the activity of design. In this way, PowerPoint acts as an "intellectual transitional" software tool. With it, we believe teachers will begin to see and embrace the possibilities of a "children as designers" model of education. Our hope here is to establish a foothold or beachhead upon which other approaches and technologies can build.

2.7 References

Aarseth, E. (2003). *Playing research: Methodological approaches to game analysis*. Melbourne, Australia: Digital Art and Culture.

Bloom, B. S., Englehart, M., Furst, E., Hill, W., & Krathwohl, D. (1956). *Taxonomy of educational objectives: The classification of educational goals. Handbook I: Cognitive domain.* New York, Toronto: Longmans, Green.

Blumenfeld, P. C., Soloway, E., Marx, R. W., Krajcik, J. S., Guzdial, M., & Palinscar, A. (1991). Motivating project-based learning: Sustaining the doing, supporting the learning. Educational Psychologist, 26(3 & 4), 369-398.

Deci, E. L., & Ryan, R. M. (1985). *Intrinsic motivation and self-determination in human* behavior. New York: Plenum Press.

Lepper, M. R., Keavney, M., & Drake, M. (1996). Intrinsic motivation and extrinsic rewards: A commentary on Cameron and Pierce's Meta-analysis. *Review of Educational Research, 66*(1), 5-32.

Clark, R. (1983). Reconsidering research on learning from media. *Review of Educational Research, 53*(4), 445-459.

Csikszentmihalyi, M. (1990). *Flow: The psychology of optimal experience*. New York: Harper & Row.

Csikszentmihalyi, M. (1996). *Creativity: Flow and the psychology of discovery and invention*. New York: Harper Collins.

Dempsey, J., Lucassen, B., Gilley, W., & Rasmussen, K. (1993-1994). Since Malone's theory of intrinsically motivating instruction: What's the score in the gaming literature? *Journal of Educational Technology Systems, 22*(2), 173-183.

Dewey, J. (1916). *Democracy and education: An introduction to the philosophy of education*. New York: Macmillan.

Duffy, T. M., & Cunningham, D. J. (1996). Constructivism: Implications for the design and delivery of instruction. In D. Jonassen (Ed.), *Handbook of research for educational communications and technology* (pp. 170-198). Washington, DC: Association for Educational Communications and Technology.

Gee, J. P. (2003). *What video games have to teach us about learning and literacy*. New York: Palgrave MacMillan.

Gee, J. P. (2005). Good video games and good learning. *Phi Kappa Phi Forum, Summer*, 33-37.

Grabinger, R. S. (1996). Rich environments for active learning. In D. Jonassen (Ed.), *Handbook of research for educational communications and technology* (pp. 665-692). Washington, DC: Association for Educational Communications and Technology.

Grant, M. (2002). Getting a grip on project-based learning: Theory, cases, and recommendations. Meridian: Middle School Computer Technology Journal, 5(1), [On-line]. Available: http://www.ncsu.edu/meridian/win2002/2514/index.html

Gredler, M. E. (2003). Games and simulations and their relationships to learning. In D. Jonassen (Ed.), *Handbook of research for educational communications and technology* (2nd ed., pp. 571-581). Mahwah, NJ: Lawrence Erlbaum Associates.

Grodal, T. (2003). Stories for eye, ear, and muscles: Video games, media, and embodied experiences. In M. J. P. Wolf & B. Perron (Eds.), *The video game theory reader* (pp. 129-156). New York: Routledge.

Harel, I., & Papert, S. (Eds.). (1991). *Constructionism*. Norwood, NJ: Ablex.

Hooper, S., & Rieber, L. P. (1995). Teaching with technology. In A. C. Ornstein (Ed.), *Teaching: Theory into practice* (pp. 154-170). Needham Heights, MA: Allyn and Bacon.

Horwitz, P., & Christie, M. A. (2000). Computer-based manipulatives for teaching scientific reasoning: An example. In M. J. Jacobson & R. B. Kozma (Eds.), *Learning the sciences of the 21st century: Research, design, and implementing advanced technology learning environments* (pp. 163-191). Hillsdale, NJ: Lawrence Erlbaum Associates.

Kafai, Y. (1994). Electronic play worlds: Children's construction of video games. In Y. Kafai & M. Resnick (Eds.), *Constructionism in practice: Rethinking the roles of technology in learning*. Mahwah, NJ: Lawrence Erlbaum Associates.

Kafai, Y. (1995). *Minds in play: Computer game design as a context for children's learning*. Hillsdale, NJ: Lawrence Erlbaum Associates.

Kafai, Y., & Harel, I. (1991). Learning through design and teaching: Exploring social and collaborative aspects of constructionism. In I. Harel & S. Papert (Eds.), *Constructionism* (pp. 85-106). Norwood, NJ: Ablex.

Kafai, Y., & Resnick, M. (Eds.). (1996). *Constructionism in practice: Designing, thinking, and learning in a digital world*. Mahwah, NJ: Lawrence Erlbaum Associates.

Kafai, Y., Ching, C., & Marshall, S. (1997). Children as designers of educational multimedia software. *Computers and Education, 29*, 117-126.

Kirriemuir, J., & McFarlane, A. (2004). Literature review in games and learning: A report for NESTA Futurelab. Retrieved September 1, 2004, from http://www.nestafuturelab.org/research/reviews/08_01.htm

Lamb, A., & Teclehaimnanot, B. (2005). A decade of WebQuests: A retrospective. In M. Orey, J. McClendon & R. M. Branch (Eds.), *Educational Media and Technology Yearbook* (Vol. 30, pp. 81-101). Westport, CT: Libraries Unlimited.

Lave, J. (1988). *Cognition in practice: Mind, mathematics, and culture in everyday life*. Cambridge: Cambridge University Press.

Olive, J. (1998). Opportunities to explore and integrate mathematics with "The Geometer's Sketchpad" in designing learning environments for developing understanding of geometry and space. In R. Lehrer & D. Chazan (Eds.), *Designing learning environments for developing understanding of geometry and space* (pp. 395-418). Mahwah, NJ: Lawrence Erlbaum Associates.

Papert, S. (1991). Situating constructionism. In I. Harel & S. Papert (Eds.), *Constructionism* (pp. 1-11). Norwood, NJ: Ablex.

Papert, S. (1993). *The children's machine: Rethinking school in the age of the computer*. New York: BasicBooks.
Prensky, M. (2001). *Digital game-based learning*. New York: McGraw-Hill.
Prensky, M. (2006). *Don't bother me mom – I'm learning!* St. Paul, MN: Paragon House.
Randel, J. M., Morris, B. A., Wetzel, C. D., & Whitehill, B. V. (1992). The effectiveness of games for educational purposes: A review of recent research. *Simulation and gaming, 23*, 261-276.
Reeves, T. C. (2005). No significant differences revisited: A historical perspective on the research informing contemporary online learning. In G. Kearsley (Ed.), *Online learning: Personal reflections on the transformation of education* (pp. 299-308). Englewood Cliffs, NJ: Educational Technology Publications.
Rieber, L. P., & Matzko, M. J. (2001). Serious design of serious play in physics. Educational Technology, 41(1), 14-24.
Rieber, L. P., Luke, N., & Smith, J. (1998). Project KID DESIGNER: Constructivism at work through play. *Meridian: Middle School Computer Technology Journal, 1*(1), [On-line]. Available http://www.ncsu.edu/meridian/jan98/index.html.
Rieber, L. P., Smith, L., & Noah, D. (1998). The value of serious play. Educational Technology, 38(6), 29-37.
Roschelle, J., Kaput, J., & Stroup, W. (2000). SimCalc: Accelerating student engagement with the mathematics of change. In M. J. Jacobson & R. B. Kozma (Eds.), *Learning the sciences of the 21st century: Research, design, and implementing advanced technology learning environments* (pp. 47-75). Hillsdale, NJ: Lawrence Erlbaum Associates.
Russell, T. L. (1997). *The "no significant difference" phenomenon as reported in 248 research reports, summaries, and papers* (4 ed.). Raleigh, NC: North Carolina State University.
Ryan, R. M., Rigby, C. S. & Przybylski, A. (2006). The motivational pull of video games: A self-determination theory approach. *Motivation and Education, 30*(4), 344-360.
Spiro, R., Kolodner, J., Pea, R., Roschelle, J., Soloway, E., Scardamalia, M., et al. (2002). *You say you want a revolution...? Can new technologies enable radically new kinds of learning? Part 1.* Paper presented at the Annual Conference of the American Educational Research Association, New Orleans.
Squire, K. (2002). Cultural framing of computer/video games. *The International Journal of Computer Game Research, 2*(1), Available online: http://www.gamestudies.org/0102/squire/
Sutton-Smith, B. (1997). *The ambiguity of play*. Cambridge, Mass: Harvard University Press.
Wineburg, S. (2001). *Historical Thinking and Other Unnatural Acts: Changing the Future of Teaching the Past*. Philadelphia, PA: Temple University Press.
Wolf, M. J. (2001). Narrative in the video game. In M. J. P. Wolf (Ed.), *The medium of the video game* (pp. 93-112). Austin, TX: University of Texas Press.

Chapter 3
Video Games, Learning, and "Content"

James Paul Gee

Abstract

In this paper, I argue both that good video games recruit good learning and that game design is inherently connected to designing good learning for players. Good game design has a lot to teach us about good learning and contemporary learning theory has something to teach us about how to design even better and deeper games. I view learning in games as a form of "experiential learning" and discuss the conditions—often met in good games—under which learning from experience is most effective for learning.

3.1 Experience and Learning

In this paper, I want to argue both that good video games recruit good learning and that game design is inherently connected to designing good learning for players (Gee 2003, 2005). Good game design has a lot to teach us about good learning and contemporary learning theory has something to teach us about how to design even better and deeper games.

Let's start with contemporary learning theory (Bransford, Brown, and Cocking 2000; Sawyer 2006). When today's learning scientists talk about the mind, it sometimes seems as if they are talking about video games. Earlier learning theory argued that the mind works like a calculating device, something like a digital computer. On this view, humans think and learn by manipulating abstract symbols via logic-like rules. Newer work, however, argues that people primarily think and learn through *experiences* they have had, not through abstract calculations and generalizations (Barsalou 1999a, b; Churchland & Sejnowski 1992; Clark 1993, 1997; Gee 1992, 2004; Hawkins 2005; Schank 1999). People store these experiences in memory—and human long-term memory is now viewed as nearly limitless—and use them to run simulations in their minds to prepare for problem solving in new situations. These simulations help them form hypotheses about how to proceed in the new situation based on past experiences.

However, things are not quite that simple. There are conditions experiences need to meet to be truly useful for learning (diSessa 2000; Gee 2004; Kolodner 1993, 1997, 2006; Shank 1982, 1999; here I follow Kolodner 2006: p. 227 most closely, though I view learning as less "text based" than does Kolodner). Since I will later argue that video games offer players experiences and recruit learning as a form of pleasure and mastery, I will argue, as well, that these conditions are properties of good games design, as well.

First, experiences are most useful for future problem solving if the experience is structured by specific goals. Humans store their experiences best in terms of goals and how these goals did or did not work out.

Second, for experiences to be useful for future problem solving, they have to be interpreted. Interpreting experience means thinking—in action and after action—about how our goals relate to our reasoning in the situation. It means, as well, extracting lessons learned and anticipating when and where those lessons might be useful.

Third, people learn best from their experiences when they get immediate feedback during those experiences so they can recognize and assess their errors and see where their expectations have failed. It is important, too, that they are encouraged to explain why their errors and expectation failures happened and what they could have done differently.

Fourth, learners need ample opportunities to apply their previous experiences—as interpreted—to new similar situations, so they can "debug" and improve their interpretations of these experiences, gradually generalizing them beyond specific contexts.

Fifth, learners need to learn from the interpreted experiences and explanations of other people, including both peers and experts. Social interaction, discussion, and sharing with peers, as well as mentoring from more advanced others, are important. Debriefing—that is, talking about why and how things worked in the accomplishment of goals—after an experience is important. Mentoring is best done through dialogue, modeling, worked examples, and certain forms of overt instruction, often "just in time" (when the learner can use it) or "on demand" (when the learner is ready).

One way to look at what is going on here is this: When the above conditions are met, people's experiences are organized, in memory, in such a way that they can draw on those experiences as a data bank to build simulations in their minds that allow them to prepare for action (Glenberg 1997; Glenberg and Robertson 1999). They can test out things in their mind before they act and they can adjust their predictions after they have acted and gotten feedback. They can play various roles in their own simulations, seeing how various goals might be accomplished, just like a gamer playing a video game. The simulations we humans run—and there are various neural accounts of how this works (Barsalou 1999a,b; Clark 1993; Hawkins 2005)—are composites of our interpreted experiences built to prepare us to predict, act, and assess. Interpreted experiences are the engine from which we build simulations.

3.2 A Piece of Research: Action, Simulation, and Reading

One interesting line of research that exemplifies these points is Art Glenberg's work (see also Schwartz & Heiser 2006 for further exemplification). Glenberg et al. (2004) describe an experiment where young children read a passage and manipulate plastic figures so that they can portray the actions and relationships in the passage. By manipulating the figures, the children get a structured embodied experience with a clear goal (portray the action in the text). After some practice doing this, the children were asked to simply imagine manipulating the figures. This is a request to engage in simulation in their heads. As a posttest, the children read a final passage without any prompting.

Children who completed the sequence of embodied experience then simulation were better at remembering and drawing inferences about the new passage compared to children who received no training. They were better compared, as well, to children who were only instructed to imagine the passage. And, most interestingly, they were better compared to children who manipulated the figures without the intermediate instructions to imagine manipulating. Encouraging simulation through the initial use of physical enactment helped the children learn a new reading comprehension strategy, namely a strategy where they called on their experiences in the world to build simulations for understanding a text in specific ways.

3.3 Social Identity and Learning

Modern learning theory tends to stress the social and cultural more than I done so far (Brown, Collins, and Dugid 1989; Gee 2004; Hutchins 1995; Lave 1996; Lave & Wenger 1991; Tomasello 1999). The reason for this is that the elements of good learning experiences—namely goals, interpretations, practice, explanations, debriefing, and feedback—have to come from some place. In fact, they usually flow from participation in, or apprenticeship to, a social group or what are sometimes called "communities of practice" (Wenger 1998; Wenger, McDermott, and Snyder 2002) or affiliation groups (Gee 2003, 2004). For instance, I am a bird watcher and I have lots of experience looking for birds. But my experiences in this domain have been greatly shaped by other people and institutions devoted to birds and bird watching.

What we might call a "social identity" is crucial for learning. For example, consider learning to be a SWAT team member. The sorts of goals one should have in a given situation; the ways in which one should interpret and assess one's experiences in those situations; the sorts of feedback one should receive and react to; the ways in which one uses specific tools and technologies, all these flow from the values, established practices, knowledge, and skills of experienced SWAT members. They all flow from the identity of being or seeking to become such a person.

What is true of being a SWAT team member is equally true of being a bird watcher, teacher, carpenter, "good" elementary school student, scientist, community activist, soccer player, gang member, or anything else. Social groups exist to induct newcomers into distinctive experiences and ways of interpreting and using those experiences for achieving goals and solving problems. Today, of course, social groups can engage in interactions at a distance via the Internet and other technological devices, so the role of face-to-face interaction is, in many cases, changing and new forms of social organization around identity are emerging.

Good learning requires participation—however vicarious—in some social group that helps learners understand and make sense of their experience in certain ways. It helps them understand the nature and purpose of the goals, interpretations, practice, explanations, debriefing, and feedback that are integral to learning.

3.4 Game Design

What's a video game? In many cases—for example, in the case of games like *SWAT4*, *Deus Ex*, *Half Life*, or *Chibi Robo*—a video game is a set of experiences a player participates in from a particular perspective, namely the perspective of the character or characters the player controls. Of course, not all games offer the player an avatar: while this fact is important, I will deal with it later, where we will see that having an avatar is just one way of achieving "micro-control", one of the defining features of video games. For the moment, I will stick with games played from a first- or third-person perspective.

Video games like those I have just mentioned are designed to set up certain goals for players, but often leave players free to achieve these goals in their own ways. The game may also allow players to construct their own goals, but only within the rule space designed into the game (for example, you can accomplish things in different ways in *Thief*, but robust hand to hand combat is not one of them). Level design ensures that players both get lots of practice applying what they have earlier experienced in similar situations (within a level) and in somewhat less similar situations (across levels). Feedback is given moment by moment and often summatively at the end of a level or in boss battles. So a number of our learning conditions are met as a matter of the basic design of the game.

Such games also encourage players to interpret their experiences in certain ways and to seek explanations for their errors and expectation failures. Such encouragement works through in-game features like the increasing degrees of difficulty that a player faces as the levels of a game advance or when facing a boss that requires rethinking what one has already learned. However, it is precisely here that talking about "games" and not "gaming" as a social practice falls short. A good deal of reflection and interpretation stems from the social settings and practices within which games are situated.

Reflection and interpretation are encouraged not just through in-game design features, but also through socially-shared practices like faqs or strategy guides, cheats, forums, and other players in and out of multiplayer settings. Gamers often organize themselves into communities of practice that create social identities ("being-doing a gamer" of a certain sort) with distinctive ways of talking, interacting, and interpreting experiences and applying values, knowledge, and skill to achieve goals and solve problems. This is a crucial point for those who wish to make so-called "serious games": to gain the sorts of desired learning effects will often require as much care about the social system (the learning system) in which they game is placed as the in-game design itself.

Because this last point is crucial, let me distinguish between what I will call the "game", with a little 'g', and the "Game", with a big "G" (Gee 1996; Shaffer 2007). The "game" is the software in the box and all the elements of in-game design. The "Game" is the social setting into which the game is placed, all the interactions that go on around the game. I will write "G/game" when I mean both together.

Both games and Games are crucial for good learning and, I would argue, good game design. But let me talk here directly about game design (in-game design), not Game design (the design of the interactions around the game). Video games offer people experiences in a virtual world (which we will see below is linked tightly to the real world) and they use learning, problem solving, and mastery for engagement and pleasure. It should be noted that humans and other primates find learning and mastery deeply, even biologically, pleasurable under the right conditions, though often not the ones they face in school (Blum 2002). Thus, I want to argue that game design is not accidentally related to learning, but, rather, that learning is integral to it. Game design is applied learning theory and good game designers have discovered important principles of learning without needing to be or become academic learning theorists.

3.5 The Situated Learning Matrix

One way in which game design and modern learning theory come together is in what I will call the "Situated Learning Matrix", which is one way to spell out what "situated learning" means. First, consider that any learning experience has some **content**, that is, some facts, principles, information, and skills that need to be mastered. So the question immediately arises as to how this content ought to be taught? Should it be the main focus of the learning and taught quite directly? Or should the content be subordinated to something else and taught via that something else? Schools usually opt for the former approach, games for the latter.

Modern learning theory (see. e.g., Bransford, Brown, and Cocking 2000; Sawyer 2006) suggest the game approach is the better one.

One version of the game approach is what I call the Situated Learning Matrix. To see what I mean by this term, let's take a concrete case, the game *SWAT4*. There is lots of content to be mastered in learning to be a SWAT team member, some of which is embedded in *SWAT4* This content involves things like how a team should form up to enter a room safely, where to position oneself in an unsafe environment, how to subdue people with guns without killing them, and facts about the range and firing power of specific weapons, ammunition, and grenades, and much else.

But the game does not start with or focus on this content, save for a tutorial that teaches just enough of it so the player can learn the rest by playing within the situated leaning matrix that is the game itself. Rather, the game focuses first and foremost on an **identity** that is, being a SWAT team member. What do I mean by calling this an "identity". I mean a "way of being in the world" that is integrally connected to two things: first, characteristic goals, namely, in this case, **goals** of the sort a SWAT team characteristically has; and, second, characteristic **norms** composed of rules or principles or guidelines by which to act and evaluate one's actions—in this case, these norms are those adopted by SWAT teams.

In some games—and this is true of *SWAT4*—the norms amount, in part, also to a value system, even a moral system (e.g., don't shoot people, even if they have a gun, until you have warned them you are a policeman). Without such norms one does not know how to act and how to evaluate the results of one's actions as good or bad, acceptable or not. Of course, norms and goals are closely related in that the norms guide how we act on our goals and assess those attempts. In a game like *SWAT4*, I am who I am (a SWAT team member) because I have certain sorts of goals and follow certain norms and values that cause me to see the world, respond to the world, and act on the world in a certain way.

To accomplish goals within norms and values, the player/learner must master a certain set of skills, facts, principles, and procedures: must gain certain sorts of **content** knowledge. However, in a game like *SWAT4*, players are not left all alone to accomplish this content mastery. Rather, they are given various tools and technologies that fit particularly well with their goals and norms and that help them master the content by using these tools and technologies in active problem solving contexts. These tools and technologies mediate between—help explicate the connection between—the players' identity (goals and norms), on the one hand, and the content the player must master on the other. The SWAT team's doorstop device is a good example. This little tool integrally connects two things: on the one hand, the team's goal of entering rooms safely and norm of doing so as non-violently as possible and, on the other, the content knowledge that going in one door with other open doors behind you can lead to being blindsided and ambushed from behind, an ambush in which both you and innocent bystanders may be killed. Let me be clear, though, what I mean by tools and technologies. I am using these terms expansively. First, in *SWAT4* tools and technologies include types of guns, ammunition, grenades, goggles, armor, lightsticks, communication devices, door stops, and so on and so forth. Second, tools and technologies also include one's fellow SWAT team members—artificially intelligent NPCs—to whom the player can issue orders and who have lots of built in knowledge and skills to carry out those orders. This allows players initially to be more competent than they are all by themselves—players can perform before they are fully competent and attain competence through performance. Further, it means that the NPCs model correct skills and knowledge for the player.

Third, tools and technologies include forms of built in collaboration with the NPCs and, in multiplayer versions of the game, forms of collaboration, participation, and interaction with real people, peers at different levels of skill. These forms of collaboration go further when the player enters web sites and chat rooms, or uses guides, as part of a community of practice built around the game. Thus, I am counting NPCS as smart tools and real people as tools, too, when we can coordinate ourselves with their knowledge and skills.

So, tools and technologies, in all these senses mediate between identity and content. rendering that content meaningful. I know why, for instance, I need to know about open doors behind me. This knowledge is not just a matter of isolated and irrelevant facts. It's a matter now of being and becoming a good SWAT team member. And I have the tool to connect the two—the doorstop.

But this mediation means, of course, that players always learn in specific contexts. That is, they learn through specific embodied experiences in the virtual world (the player has a bodily presence in the game through the character or characters he or she controls). And, indeed, one hears a lot these days about learning in context. However, contexts in a game like *SWAT4* are special. While they are richly detailed and specific, they are, in reality, not just any old contexts, but richly *designed problem spaces* containing problems that fall into a set of similar, but varied challenges across the levels of the game.

Context here, then, means a *goal-driven problem space*. As players move through contexts—each containing similar but varied problems—this helps them to interpret and eventually generalize their experiences. They learn to generalize—but always with appropriate customization for specific different contexts—their skills, procedures, principles, and use of information. This essentially solves the dilemma that learning in context can leave learners with knowledge that is too context specific (because they have too few experiences and paid too much attention to their specific details), but that learning out of context leaves learners with knowledge they cannot apply (because learners have no experiences). Players come to see specific solutions as members of more general types of approaches, for example, "rushing" (building and attacking fast) versus "turtling" (building and attacking more slowly and methodically) in a real-time-strategy game).

Below I give a diagrammatic representation of the Situated Learning Matrix. The Matrix has "Experience" on top, with the conditions that render experience efficacious for learning. The material inside the box is meant to show how identity is formed from goals and norms, goals and norms which are exemplified in tools and technologies. These tools and technologies are, in turn, rendered and render contexts meaningful as particular types of problem spaces seen in specific ways. Finally, these problem-solving contexts are what render facts and information ("content") meaningful and memorable. "Content" is recruited by problems, tools and technologies, goals and norms, and identities for uses (functions) and that is what gives content meaning (makes it content in the first place).

Good games—together with their associated Games—ensure that these conditions are met, but in an educational context, things need to go further here. This is where "teachers" and mentors (not necessarily official teachers in the school sense) become crucial. Such teachers and mentors help create a Game that ensures that experiences will be well interrogated in the game.

SITUATED LEARNING MATRIX

Experience:
Goals, Interpretation, Feedback, Explanation, Practice,
Social Interaction (mentoring, sharing, debriefing)

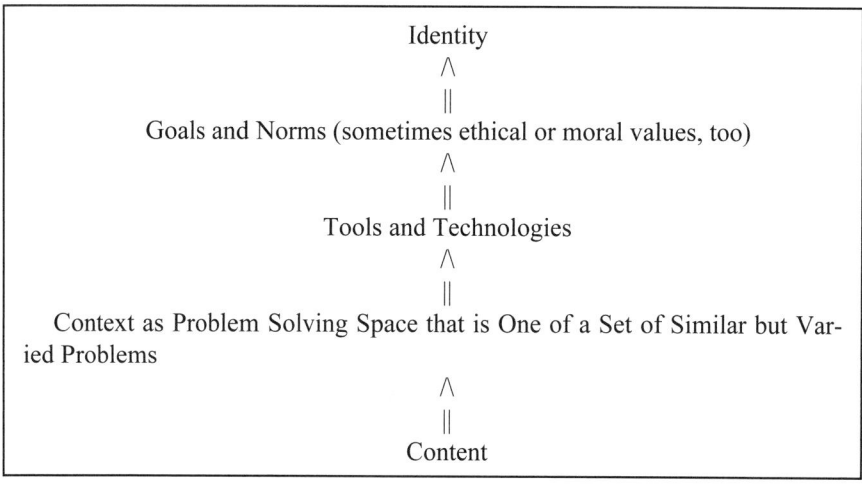

Why do I call this a SITUATED learning matrix? Because content is rooted in experiences a person is having as part and parcel of taking on a specific identity (in terms of goals and norms stemming from a social group, like SWAT team members). Learning is situated in experience, but goal-driven, identity-focused experience. This is one way to look at what situated learning theory is all about.

Learning in video games—learning in terms of the Situated Learning Matrix—is not "anything goes", "just turn the learners loose to do their own thing" (Kirschner, Sweller, & Clark 2006). There is a good deal of guidance in games. Guidance from the game design itself, from the NPCs and the environment, from information given "just in time" and "on demand", from other players in and out of the game, and from the resources of communities of practice built up around the game.

3.6 Clearing Up Possible Misconceptions

By using *SWAT4* as my example, I may well have created a number of misconceptions. So, let me try to clear them up. All games have content, that is, facts and skills one must master. This content is crystal clear in *SWAT4* because the game is organized around a connection to a real-world domain. But consider a game like *Chibi-Robo*. In this game, one takes on the identity of a four-inch house-cleaning robot enjoined to make people happy. This identity defines and, in turn, flows from one's goals in the game (e.g., solve problems in the house by cleaning and getting to hard-to-reach places where you can do or get something that will make people happy) and norms (e.g., you should do good, not bad, for the people and animated toys in the house; it is important to repeatedly talk to everyone; problems must be solved in a time efficient way, since your battery only lasts so long). There are lots of skills and facts one needs to master, all of them germane to being a four-inch robot, skills such as how to get to high places and facts like: falling from such high places will take a good deal off your allowable time to solve problems. And there are tools that connect your identity and theses skills and facts, for example, a little rotor that you can use to soften your fall.

The same can be said of *Half-Life*, *Deus Ex*, *Civilization*, or any other such game. The skills and facts in these games are content, but usually not recognized as such unless they fall into a real-world domain like physics or SWAT teams.

Finally, using *SWAT4* as an example, may lead to the conception that the Situated Learning Matrix is only germane to violence or, at least, to action filled adventures. It can seem that content—knowledge—in a domain like science or art is much less connected to identities, goals, and values. However, ethnographic accounts of scientists learning and doing science, for instance, show this is not true Latour & Woolgar 1979; Ochs, Gonzales, Jacoby 1996; Traweek 1988). And, for students in school, there is clear research that shows that content divorced from the Situated Learning Matrix is inert and unable to be applied in practice, however much the student may pass multiple choice tests (Chi, Feltovich, & Glaser 1981; Gardner 1991).

3.7 References

Barsalou, L. W. (1999a). Language comprehension: Archival memory or preparation for situated action. *Discourse Processes* 28: 61-80.
Barsalou, L. W. (1999b). Perceptual symbol systems. *Behavioral and Brain Sciences* 22: 577-660.

Bransford, J., Brown, A. L., and Cocking, R. R., (2000). *How people learn: Brain, mind, experience, and school: Expanded Edition.* Washington, DC: National Academy Press.

Brown, J. S., Collins, A., & Dugid (1989). Situated cognition and the culture of learning. *Educational Researcher* 18: 32-42.

Chi, M.T.H., Feltovich, P. J., & Glaser, R. (1981). Categorization and representation of physics problems by experts and novices. *Cognitive Science* 13: 145-182.

Churchland, P. S. & Sejnowski, T. J. (1992). *The computational brain.* Cambridge, Mass.: Bradford/MIT Press.

Clark, A. (1993). *Associative engines: Connectionism, concepts, and representational change.* Cambridge: Cambridge University Press.

Clark, A. (1997). *Being there: Putting brain, body, and world together again.* Cambridge, Mass.: MIT Press.

diSessa, A. A. (2000). *Changing minds: Computers, learning, and literacy.* Cambridge, Mass.: MIT Press.

diSessa, A. A. (2004). Metarepresentation: Native competence and targets for instruction. *Cognition and Instruction* 22: 293-331.

Gardner, H. (1991). *The unschooled mind: How children think and how schools should teach.* New York: Basic Books.

Gee, J. P. (1992). *The social mind: Language, ideology, and social practice.* New York: Bergin & Garvey.

Gee, J. P. (1996) *Social linguistics and literacies: Ideology in Discourses.* London: Taylor & Francis (First Edition, 1990).

Gee, J. P. (2003). *What video games have to teach us about learning and literacy.* New York: Palgrave/Macmillan.

Gee, J. P. (2004). *Situated language and learning: A Critique of traditional schooling.* London: Routledge.

Gee, J. P. (2005). *Why video games are good for your soul: Pleasure and learning.* Melbourne: Common Ground.

Glenberg, A. M. (1997). What is memory for. *Behavioral and Brain Sciences* 20: 1-55.

Glenberg, A. M., Gutierrez, T., Levin, J. R., Japuntich, S., & Kaschak, M. P. (2004). Activity and imagined activity can enhance young children's reading comprehension. *Journal of Educational Psychology* 96: 424-436.

Glenberg, A. M. & Robertson, D. A. (1999). Indexical understanding of instructions. *Discourse Processes* 28: 1-26.

Hawkins, J. (2005). *On intelligence.* New York: Henry Holt.

Hutchins, E. (1995). *Cognition in the wild.* Cambridge, MA.: MIT Press.

Kirschner, P. A., Sweller, J., & Clark, R. E. (2006). Why Minimal Guidance during Instruction Does Not Work: An Analysis of the Failure of Constructivist, Discovery, Problem-Based, Experiential, and Inquiry-Based Teaching. *Educational Psychologist* 41: 75-86.

Kolodner, J. L. (1993). *Case based reasoning.* San Mateo, CA: Morgan Kaufmann Publishers.

Kolodner, J. L. (1997). Educational implications of analogy: A view from case-based reasoning. *American Psychologist* 52: 57-66.

Kolodner, J. L. (2006). Case-based reasoning. In R. K. Sawyer, Ed., *The Cambridge handbook of the learning sciences.* Cambridge: Cambridge University Press, pp. 225-242.

Latour, B. and Woolgar, S. (1979). *Laboratory life: the social construction of scientific facts.* Los Angeles, CA: Sage.

Lave, J. (1996). Teaching, as learning, in practice. *Mind, Culture, and Activity* 3: 149-164.

Lave, J. & Wenger, E. (1991). *Situated learning: Legitimate peripheral participation.* New York: Cambridge University Press.

Ochs, E., Gonzales, P. & Jacoby, S. (1996). "When I come down I'm in the domain state." In E. Ochs, E. Schegloff, & S. A. Thompson, Eds., *Interaction and Grammar.* Cambridge: Cambridge University Press, 328-369.

Sawyer, R. K., Ed. (2006). *The Cambridge handbook of the learning sciences.* Cambridge: Cambridge University Press.
Shaffer, D. W. (2007). *How computer games help children learn.* New York: Palgrave/Macmillan.
Schank, R. C. (1982). *Dynamic memory.* New York: Cambridge University Press.
Schank, R. C. (1999). *Dynamic memory revisited.* New York: Cambridge University Press.
Schwartz, D. L. & Heiser, J. (2006). Spatial representations and imagery in learning. In R. K. Sawyer, Ed., *The Cambridge handbook of the learning sciences.* Cambridge: Cambridge University Press, pp. 283-298.
Tomasello, M. (1999). *The cultural origins of human cognition.* Cambridge, MA: Harvard University Press.
Traweek, S. (1988). *Beamtimes and lifetimes: The world of high-energy physicists.* Cambridge, MA: Harvard University Press.
Wenger, E. (1998). *Communities of practice: Learning, meaning, and identity.* Cambridge: Cambridge University Press.
Wenger, E., McDermott, R., & Snyder, W. M. (2002). *Cultivating communities of practice.* Cambridge, MA: Harvard Business School.

Chapter 4
Fair Game:

Gender Differences in Educational Games

Kimberely Fletcher Nettleton

Abstract

Historically, toys and games have been provided for boys and girls. While boys and girls play together, and often play the same games, their enjoyment in the games is often quite different. Looking at these games can shed light on how the minds of each gender are engaged while playing.

    Educationally, games are used as a vehicle to engage students in the learning process. They are used to drill facts, connect ideas, or help students synthesize discrete knowledge. How each gender uses games in the learning process can have an impact on the learning which will take place. Do current educational games, because of their design, cause a gender gap in learning? As schools use more and more games as part of the curriculum, will there be a wider divide in the learning which takes place?

    This chapter will examine the differences between the genders in play and learning. It will examine games, used historically and currently, that appeal to boys and girls. It will examine at what age the differences in learning styles between genders begins to appear and when they become most pronounced. The chapter will focus on what the current research suggests about the way each gender learns and retains knowledge. Finally, the chapter will address learning style differences and raise questions concerning game design and its impact on gender-based learning.

4.1 Introduction

A small boy in Ancient Greece rolls a clay marble into a circle drawn in the dirt… a little girl in medieval England sings lullabies to her cloth poppet… a Pharaoh's son pushes a metal chariot across the floor… parents have been giving their children toys for centuries. Dolls, rattles, toy horses, tops, marbles, and child sized weapons have been found at many archeological sites throughout the world. Toys have been a hot commodity in England as far back as 1300. From tea sets to toy soldiers, toys were produced and sold (Orme, 2001).

Historically, toys and games have been provided for children for either amusement or to develop skills. Many games and toys are designed to enhance recreation. There have always been toys, however, that were used to teach. It has been postulated that many early playthings were developed to reinforce skills that children needed to thrive in society. For example, a child sized bow and arrow may not have been provided as much for play as for learning hunting skills. In a culture where tools were made by hand and fire came from friction, a twirling stick toy is thought to have been used to reinforce the drilling skills necessary for survival (Elkonian, 1999). Before formal education, toys were used to train children for the skills they would need to succeed in life.

Modern parents buy toys for their children for some of the same reasons they provided for children in the past. While there are many toys that are designed for leisure, there is a plethora of toys that are designed to teach. Modern parents are just as concerned about their children surviving in the asphalt jungle as prehistoric parents were about their children surviving in the primeval jungle. Today's parents want to ensure that their children have developed an academic edge by the time they hit preschool. Contemporary parents are purchasing truckloads of the many educational toys flooding the market. Educational toys seem to be new phenomena, but who knows how many toys would have been bartered for in primitive societies if slick merchandising had been invented? Whether toys are chosen because they are reinforcing skills or because they are touching a chord somewhere deep within the psyche of children, many toys for boys and girls have remained relatively unchanged for thousands of years (Elkonian, 1999).

There is a difference in the style of games, play, and toys that boys and girls have traditionally used. The reasons for these differences may be due to innate differences in the brain or stereotypical conditioning on the part of parents and society. No matter why they exist, they are a part of our current culture and need to be examined. Games can be a powerful educational tool and need to be finely crafted to provide curriculum content. In order to be effective, they must not only improve weak skills, but enhance competencies of boys and girls.

4.2 Gender Differences in Toys and Games

From a very young age, boys and girls are provided with different toys. A study of toys provided for children between the ages of one and six found that "boys were provided with more toys of more classes (depots, educational-art materials, machines, military toys, spatial-temporal objects, sports equipment, toy animals, and vehicles) than the girls (dolls, "doll houses", and domestic objects)" (Rheingold & Cook, 1975, p. 462). In spite of the fact that Rheingold and Cook (1975) observed both eighteen month old boys and girls playing with trucks for the same amount of time, the bedrooms of those girls did not contain any vehicles. Children at this young age do not choose their toys, their parents do. These decisions are based on gender.

Visit any toy store and it will be divided into different colored sections: The bright, primary colors are the baby and toddler aisle. These toys are designed to appeal to both sexes by their bright colors and the sounds they make. The dark aisle is where boys' toys are: action figures, cars, and an arsenal of weapons designed to save the galaxy. The pink aisle is next, filled with girl toys: Barbies, tea sets, and tiaras. If girls are going to save the world, they are going to look good doing so. Finally, there is the rainbow aisle of games, an eclectic aisle, in which it is usually safe to buy for either gender. While a few are packaged specifically for one gender or the other, there are many games that are enjoyed by both girls and boys. Many popular board games appear to have become almost gender neutral. Early electronic games, such as Pong, appealed to both sexes (Sweedyk & de Lael, 2005).

Toys and games are a connection between the past and present. Games from several continents: Europe, Asia, Africa and South America, are still played after being a part of the culture for hundreds of years. "Children in the Arctic play string games which have been passed down for centuries" (Bell, 1979, p. 136). One of the earliest game boards found by archeologists came from royal tombs in the city of Ur. Dating from around 3000 B.C.E., "the game board was made of baked clay and looks somewhat like a cribbage board" (Bell, 1979, p. 24). It is unclear how this game was played, but it was obviously important enough to be interred. Dogs and Jackals, a board game found in a tomb in Thebes, dates back to almost 2000 B.C.E. (Bell, 1979). Perhaps the oldest known board game is Senet. Several elaborately carved playing pieces and boards have survived. Pictured on both papyri and the walls of tombs in Egypt, it was very popular. So popular, in fact, that game directions were never provided. It was assumed that everyone knew how to play (Bell, 1979).

Games have been played all over the world. Some games have traditionally been played by women only. One of those games is the "walnut shell game, played by the women of the Yokut tribe in North America" (Bell, 1979, p. 80). Women's games were usually an outgrowth of a tedious, communal chore. They made the time pass more quickly. In Africa, the board game, Mancala, has been around for thousands of years. Mancala boards have been made of a variety of materials: clay, wood, or ivory. An impromptu board could easily be made by digging small holes in the dirt and collecting seeds, beans, or small stones for play. There are many variations of the game and it is still played today. Mancala was "considered a man's game, although women often play. If a woman became a strong player, men refused to play with her. The game was played for prestige, and no man's reputation would bear the ridicule of being defeated by a woman" (Bell, 1979, p. 121).

Games have not only been developed to pass the time or to provide social status. They have also been designed to teach. Monopoly, one of the most popular board games sold in the last 70 years, was originally created to teach about land monopolies. The inventor, Lizzie Magie, used the monopoly board she developed to teach students about land tax theories. The game was used as a teaching tool, but was revamped for sale by Parker Brothers (Management Today, 2007).

A careful examination of the way games are played, how often they are played, and how much enjoyment is achieved through playing, shows that there are differences between the genders. It is not enough to just examine the methods each gender employs when playing games, it is important to look at what satisfaction children receive from playing. While boys and girls play together (and often play the same games, separately), their level of involvement in the games can be different.

A game is very structured, with rules spelled out very clearly. These rules may evolve over time, but the variations do not change the basic structure of the game. Basketball added the three second rule, the three point line, and the shot clock, but the game being played on the floor is still, fundamentally, basketball. Many researchers make a distinction between the terms game and play. Play is considered to be loosely structured. The rules are mutually agreed upon and understood, but they are very flexible. In order to fully understand how best to design educational games for both males and females, it is important to examine how play integrates into games for girls and boys.

4.3 Fantasy Play, Simulations, and Games

Almost every little girl has played house. Role playing occurs everywhere, from the playground to the playroom. The identification of character for girls is a very important aspect when playing. Many will hang up pictures in their house, from magazines, of what they look like. They enjoy dressing up and using props to help further the fantasy: old hats and jewelry, long evening dresses, old high heels. June Cleaver may have worn pearls when cleaning, but most little girls prefer to play house decked out in satin and lace

Barbie, is another favorite past time of young girls. It's often the number one gift requested for young girls at Christmas (Otnes, Kim & Kim, 1994). Character identification with Barbie, comes from pretending that the clothes and accessories are part of a personal wardrobe. Getting Barbie, dressed for different activities is an important ritual during play (Kuther & McDonald, 2004). Girls become the character they create. This is much like the opening segments in many video games, when characters are created before play begins. The difference is that in video or electronic games, there is a proscribed path that characters must follow. In Barbie, the character is free to go in any direction.

The stories created around Barbie and her friends are part of the play. Young girls often script out entire scenarios for Barbie (Kuther & McDonald, 2004). When they play together, Barbie becomes a catalyst for talking. Play is often suspended for important talks. Girls will chat about whatever is on their minds, share information, and concerns. Then the play period resumes, as if the action was never paused. The story continues.

For many girls, the actual playing of Barbie, is secondary to setting up the house. If a girl doesn't happen to own the Barbie Dream House, then she has to create a home for Barbie. The home is the stage where all the action will take place. Putting together a home can last for hours while girls find scraps of cloth for bedspreads, boxes to make into furniture, and odd items to make into household accessories. For many, this act of creation is actually the play. Often, after so much time creating and decorating Barbie's home, the interest in actually playing with Barbie, wears off.

It may be that as girls grow older, their venues of play, changes. For some young women, creating a "look" with hair, clothes, and make up, becomes play and it is fun. While it is serious play, for many girls, shopping becomes a pastime. Indeed, "shopping centers (malls) were initially planned with the female consumer in mind" (Cohen, as cited in Scanlon, 2000). Over the years, hanging out at the mall has become a popular pastime. Teenage girls will try on clothes and accessories, even when they have no intention of making a purchase. In effect, they have become Barbie.

Instead of playing with Barbie or developing a character in an electronic game, a teenage girl can recreate herself through clothes. Shopping takes on the same aspects as play did when they were younger. Many older women, who loved to shop when young, often find it a nuisance when they are older. Has it lost its appeal because it no longer is fantasy play... there is no expectation of re-creation? Perhaps girls prefer play to games because while it is different than a goal oriented game, it appeals to them at the same emotional level as games appeal to boys. Play involves contemplation, thinking, planning and using the imagination to create. Games require a more competitive mindset, where the creativity comes in trying to reach a goal before an opponent. It may be that a simulation is the bridge between games and fantasy play.

One of the most popular, best selling electronic games, The Sims, "is one of those rare computer games played by more women than men." (Grossman & Song, 2002, para. 2). For many women, the parallel between playing The Sims and Barbie, is very strong. Creating characters, furnishing homes, shopping, and setting up situations can be very absorbing. MTV produced a documentary which explored how the lives of people playing the Sims game, mirrored their Sims creations (Anderson, 2005). There is no clear goal or end to the Sims game. A player does not play to reach the end of the game in order to win. As Grossman and Song wrote, "Some games ... you don't play to win. You just play to play." (2002, para. 10). A careful examination of the game shows that for many women, the game is a more advanced version of Barbie. After spending hours developing characters and homes, when playing the Sims, many players lose interest and create new characters. Asked if they enjoy the game, they are enthusiastic. However for some, boredom sets in and killing off the characters is not enough. Electronic

Gaming Monthly offers information on "1001 Ways to Torture Your Sims" (The Sims, 2003). Both male and female Barbie players have reported that they often engaged in torture play. From melting her in the microwave, to using her to release pent up anger or frustration, Barbie and her friend Ken often lost limbs in the course of play (Kuther & McDonald, 2004). Violence and aggression can appear even in such seemingly innocuous games.

While girls are busy creating homes, creating stories, and re-creating themselves, boys are doing some of their own creating, too. Starting as toddlers, stumbling around in their daddy's over-large shoes, boys go on to identify with superheroes. A couple of pins, an old towel, and boys are racing to save the world from evil marauders and their minions. As any mother can attest, boys are capable of making a weapon out of anything. Although close to playing House or Barbie, superhero play is actually not as complex. The superhero character usually comes complete with a proscribed set of superhero powers, so there is little need to spend time on character development. Sophisticated story lines might emerge, but they are very formula driven: good guy captures bad guy and saves the world. Boys will often brag about their exploits after they have played.

Until recently, the only dress-up, imaginary clothes for boys consisted of cowboy hats and superhero capes. In Disney World, a special boutique caters to little girls and their princess fantasies. Girls can have their hair styled, have their faces made up, and purchase full royal regalia in order to have a picture taken of themselves as a princess. Boys can visit an adventure shop, which is very dark and filled with pirate merchandise. Do boys have a secret desire to dress up as a Prince or other Hero? Is it gender bias that keeps boys relegated to the adventure shops and dark colors? Or is there an innate reason for how society treats each gender?

Young boys like to dress up in costumes and have a fascination for play jewels, rocks, and collections. There are very few games or modes of play that tap into this desire. Superhero capes and treasure boxes help. While today it is hard to imagine that boys have the desire to dress up, the clothing of the nobles during the Renaissance suggests that men enjoyed costuming every bit as much as women.

During an observation of a kindergarten class's free play period, many of the girls loved playing Pretty, Pretty, Princess. Packaged for girls, the game consists of moving around a board and collecting rings, beaded necklaces, and flashy jewelry. Attracted by the glitz, one of the little boys, Joe, announced that he wanted to play.

"That's a girls' game," another boy, Mike, told him scathingly.

"So?" said Joe, and he sat down and began playing,

Within 2 weeks the kindergarten teacher reported that all of the boys in the class, including Mike, were playing Pretty, Pretty, Princess and the girls had to elbow their way in to join the game. The jewels attracted the boys, as well as the chance to wear them. Even in kindergarten, stereotypes are strong. When one boy broke the barrier, the others were willing to follow along, lured by the glittering jewels.

Breaking the gender barrier, however, does not always happen so easily. Kuther and McDonald believe that girls are much more likely than boys to cross the gender divide. Many young men are reluctant to admit they played Barbie as children. If they confess to playing, they either blame it on the influence of a bossy sister or brag about how they had tortured Barbie (2004).

With the introduction of GI Joe, by Hasbro in 1964, boys were provided with a socially acceptable action doll. The doll was the same size as Barbie. Although popular when it was introduced, the rise of action figures more effectively changed the way boys played. Released in the spring of 1978, the Star Wars action figures were only 3 3/4 inches tall with clothes molded to the body. While they were not the first action figures on the market, they were the first to become widely popular (Leach, 2007). Subsequent action figure collections were based on television shows, cartoons, movies, comic books, or even Bible stories. The action figures were provided with antagonists as well as accessories to replicate scenes. Many action figures have rules about what it can and cannot do: the figures come with their own powers. For a small price, boys can build a whole army of action figures. Even this imaginative play, for boys, has rules that help define the structure of the game. Boys who try to make their character act in a way not already proscribed, often become embroiled in an argument with their playmates. Unfortunately, the ages at which boys can play with action figures, without ridicule by peers, is very low. Action figures rarely come to school to be played with at recess.

4.4 The Playground

When the bell for recess sounds, girls and boys run to their corners of the playground and begin their games. On almost every playground, children are separated across the gender divide. There are exceptions: the swing set, the soccer field, the four square box. For the most part, girls and boys separate to play different games (Thorne, 1993).

The chants and rhythms of little girls playing jump rope together fill the playground. They line up to twirl the rope and wait in line for their turn, when everyone will chant, count, and congratulate each other for jumping in and out, reach a high number, or completing any number of intricate steps. This game calls for good reflexes, skill, and a need to work together.

Over on another part of the playground, little boys are trying to scale a large hill of plowed snow, pushing each other back down, playing King of the Mountain. The main goal of this game is to be the King at the top and defend the mountain from others trying to capture the summit. Occasionally the boys work in teams, but this game is usually dependent on the skill of the King to push the others back down the hill. The rules are very simple and winning depends on strength and skill. This game calls for good reflexes, skill, and a desire to take over.

If both games call for skill and good reflexes, why do more boys play King of the Mountain instead of jump rope? Why is there a gender divide on the playground? Observing children on the playground shows some interesting differences between boys and girls. Whether these are based on social conditioning or innate differences, there are clear preferences in play for each sex. Most boys are very involved in team ball games. Their games usually last the entire recess periods. Ball games have clear rules that are followed by everyone. While girls play during recess, they tend to be involved with more than one activity and prefer more sedentary type games. It is not clear why girls prefer less active games during recess, but when there is equipment and some choices offered for ball games that are non-aggressive, many girls will choose to play. It may be that the choices that have been traditionally offered to them are not appealing (Blatchford, Baines, & Pelligrini, 2003). Girls usually play hopscotch, jump rope, and rhythmic games.

While boys seem to prefer games with a goal, they also engage in role playing on the playground. They will act scenes or behave like superhero characters from television or electronic games. While girls prefer games that have a language component (Pelligrini et al., 2004), they also refer to popular media characters in their play. Educators can use this type of play to enhance literacy skills and teaching strategies in the classroom (Grugeon, 2005). The story lines, characterizations, language and vocabulary inherent in media role play games can be very powerful motivational tools. Educational game designers can build on the way children naturally use popular culture role play during recess to design literacy based games.

As society changes, the playground is taking on a new look. Community soccer leagues have spawned a new playground game. Girls and boys are beginning to play soccer and other pickup ball games during recess. Girls make up a minority in these games, but they are accepted as players (Blatchford, Baines, & Pelligrini, 2003). It is often hard to make broad assumptions of gender preferences only on what is currently being observed. Careful analysis of playground interactions and games can be very helpful in designing educational games. The skills involved in ball games: eye-hand coordination, achieving a score, and reaching a goal have been easily adapted in many electronic games. The skills that girls use in their preferred playground games are either rhythmic, have a story line or are word games (Pelligrini et al., 2004).

4.5 Games

As boys grow older, they begin to exchange unstructured play for games. Boys appear to do this at a much earlier age than girls, but it could be that there have not been many venues for boys to fulfill role playing, imaginative type play. There are few packaged games that filled this void for boys (not every group of boys would be as willing to break the gender barrier and play Pretty, Pretty, Princess).

A change occurred when video games hit the market. They provided a cross-over between fantasy play and games. Like action figures, the characters of many video type games are created to interact through a series of elaborate rules. Building a character and identifying with it becomes part of the play. Gee (2003, chap. 3), notes that he deeply identified with his created character on several levels. This character engagement helps boys continue playing in a socially acceptable form.

Character identification for boys in many video games may involve putting together a character. Initially, boys tend to build stronger characters than girls. They search for strong attributes and during the game, look for ways to develop the skills of their character. On the whole, girls initially create characters that try to solve problems without resorting to violence (Hayes, 2005). Unfortunately, the current sophistication level of most video games cannot utilize girls' greatest weapons: communication and imagination. So far, games which allow girls to talk their way out of problems or to exchange meaningful communication with their nemesis, have yet to be fully realized. Currently, a character can't earn points for making a cutting remark or avoiding a confrontation. Characters don't earn points for creative solutions. As girls play role-playing video games over and over, they realize that the skills they chose for their characters are not ideal. They need to be changed to match the constructs of the game. Their characters have to adapt and develop fighting skills in order to win. Girls will change their character to adapt to the confines of the game (Hayes, 2005). While it is clear that many girls enjoy creating more aggressive characters from the start, developing a strong fighter is usually not the initial choice for many (Hayes, 2005).

Video games also provide players with a structured game format. Boys appear to have a need for reaching a goal rather than playing. Video games have set rules, as do other games that boys enjoy: baseball, football, or board games. Clarifying these rules often extends a recess into the classroom (Blatchford, 2003). It is a point of honor to fight to the death over the exact nature of a game's rule. This sophisticated negotiation often masquerades as "out shouting your opponent". Young boys spend a lot of their time arguing and establishing the rules of play. They play games where there is a clearly defined winner and loser. Games of war, action heroes, and even "cars" all have to have a strategic element to them. Video games have traditionally been the province of boys, whether through marketing blunders, or unconsciously designed appeal.

Part of the appeal for the current crop of games for boys, are the rules involved within characters. The character has to play within a structure, established by its strengths and weaknesses. Increasing the strengths of the character is part of the challenge. This happens as a character moves through game levels and acquires special strengths and skills. Building a character does not happen quickly.

One of the arguments that often arise about girls and video games is that girls are not willing to devote as much time to playing as boys. According to one study, 10th grade girls only spend 6% of their time on games, while boys spend 38%. (Canadian Teacher's Federation, 2003; Sanford &Madill, 2006). Many video games have elaborate rules and intricate skill levels. Games may not be popular with girls at this age because of the amount of time involved. In a study by Bonanno & Kommers (2005) found that the "amount of time spent on games by both men and women between their first and second years of higher education decreased, although males still spent almost 6 hours each week playing games, compared to females, who played a little more than 2 hours" (Bonanno & Kommers, 2005, 26). In fact," traditional games are the most popular among most college students, rather than imaginative games" (Bonanno & Kommers, 2005, p. 17). The games do not demand as much time and the rules are familiar, making them a good fit with time crunched students (Bonanno & Kommers, 2005).

This same time crunch may have something to do with the rise in a new genre of games called casual games. These games are designed to be "relaxing or soothing instead of energizing or exciting. The most successful games are ones that don't require too much concentration." (Moltenbrey, 2006, p. 44). Some of the most popular games of this genre include Bejeweled, Tetris, and Chuzzles. In an interview, Jason Kapalka, chief creative officer of Pop Cap Games, identified the market for casual games. The "typical casual gamer is a 40-something woman. At Pop Cap, 72 percent of our seven million monthly visitors are female, and fully three-quarters are over the age of 35. This is reflective of the industry as a whole." (Moltenbrey, 2006, p. 45).

Although the casual game market is flooded by older women, young women are also attracted to these types of games. A study of favorite games of 16-18 year old young women, found that women preferred puzzle games like solitaire over fighting games. One reason for this may be that "females prefer a more concrete, contextualized, and repetitive activity that does not demand risk taking. They prefer adopting a tinkering approach that requires rapid access and retrieval of information from memory involving comparisons and rhythmic movement." (Bonanno & Kommers, 2005, p. 29). Fighting games also require using short term memory and a learned, repetitive movement (Bonanno & Kommers, 2005). These games appeal to many young women because, like casual games, they do not have to concentrate solely on the game. Multi-tasking becomes much easier with this type of game.

At the same time, young men between the ages of 16-18 are enjoying different types of games: "first person shooter games, role playing games, sports games, and strategy games." (Bonanno & Kommers, 2005. p. 29). This gender divide may appear when choosing favorite games because "males prefer command structures that makes them feel in control, especially by continually intervening through actions guided by their prominent visual spatial capabilities and manipulating information (guessing distances, calculating angles, deciding strength of action, and so on) in working memory." (Bonanno & Kommers, 2005, p. 29). These strengths are considered to be male strengths and many games are designed to take advantage of these skills. Are these stereotyped skills or is there a scientific basis for what is perceived as gender differences between males and females?

4.6 Gender and Education

Although people clearly have individual preferences, there is a large body of work (Blatchford, Baines,& Pelligrini, 2003; Gurian, Henley, & Truman, 2001; Honigfield & Dunn, 2003; Jordan, 1995; Shalin, 1998; Thorne, 1993), which quite strongly suggests that there are differences between genders when it comes to learning, thinking, and playing. These differences are seen across cultures. It is important to differentiate between bias, stereotype, and true gender differences. Individual differences can be stronger than gender differences (Honigsfeld & Dunn, 2003), but to really understand gender differences, it is important to understand how the brain works. Brain research has been enhanced by using magnetic resonance imaging (MRI). Researchers can look closely at the development of the brain and the way it works (Gurian, Henley, & Trueman. 2001).

Even before birth, the brain is subjected to hormonal influences (Gurian, Henley, & Trueman. 2001). Girls and boys are different when it comes to collecting, processing, and using information. With more neuronal fibers, girls' brains are wired to help them focus on subtle changes and differences in the emotions of others (Shalin, 1998). Girls have more active frontal lobes and faster maturing language centers. Girls have a greater facility for language. Their verbal ability is considered to be greater. Girls tend to interact more with others through language. Overall, they are better at multi-tasking (Gurian, Henley, & Trueman, 2001).

Boys are visually oriented. Their spatial abilities appear to be much more advanced than most girls. Boys tend to be better at abstract reasoning. They tend to be better at storing trivia for a longer period of time than girls (Gurian, Henley, & Trueman, 2001). This trait certainly gives boys an edge in video games. Boys will talk for hours about the skill levels of video game characters, juggling details, and by weighing their relative merits, determine which character is the strongest. Over all, boys are more aggressive than girls and tend to be more goal oriented. They tend to become absorbed in something and focus on one thing at a time (Gurian, Henley, & Trueman. 2001). It is these differences that are at work when boys and girls play games and learn.

Sometimes it is hard to differentiate between stereotype, cultural bias, and true innate differences between genders, because there is a myriad of individual differences between each gender. Yet, even across cultures, disparities among the genders have emerged. It is these differences that need to be carefully scrutinized for educational purposes.

When investigating the learning styles of adolescents from eight different countries around the world, some consistent variations between boys and girls were revealed. Boys were found to be more kinesthetic and peer oriented. This means boys learn best by doing and working with others (Honigsfeld & Dunn, 2003). While culture and society may influence how boys and girls behave, what they prefer, and how they act, the fact that gender differences can be found in children around the world helps support the idea that some dissimilarities are innate.

Gender based education has been designed to teach to these differences. Although brain research has some identified some gender differences in the structure and activity levels in the brain, there are many differences between girls and between boys that make every person unique. Every person is an individual and it is important to acknowledge that gender based education is not a one size fits all proposition. Many schools are experimenting with all boy or all girl classrooms, designed to teach to the strengths of each sex. However, since wide variations occur between boys and between girls, many schools are keeping students together in co-ed classrooms, but are training teachers to teach to the strengths of both sexes (Erlauer-Myrah, 2006). Even with emerging brain research, the mystery of how learning takes place is not fully understood. Many educators believe that students learn by connecting new experiences to previous ones (Taylor, 2006). This scaffolding helps anchor knowledge in the minds of learners.

4.7 Games in the Classroom

Helping people connect new information to old is not an easy task. Nor is it easy to connect context to curriculum (Park, 2006). Games are a way of helping make the link. Educationally, games are used as a vehicle to engage students in the learning process. They are used to drill facts, connect ideas, or help students synthesize discrete knowledge.

Games serve many purposes in the classroom. They are often used as time fillers, with no specific educational value. For example, 7 Up, or Huckle Buckle Bean Stalk are often played at the end of a class period or the end of the day, when there are just a few minutes to spare. Popular classroom games that are used to add interest to review sessions include Around the World, Bingo, and some form of Jeopardy. Many other learning games are created by teachers. Educators often make file folder games. These are usually simple board games with question cards that reinforce subject matter. While these are not especially appealing to either gender, they serve a purpose by helping students review facts. Some games teach skills or content. The Binary Game was designed to teach skills about Binary numbers (McLean, 2006), while many teachers use Battle Ship to teach coordinate graphing skills. Whether a game teaches a skill or reinforces it, a game can be a powerful tool for educators and students.

One method of teaching, used by many educators, is the Team-Games-Tournament (T-G-T) model. Students are divided into base teams and then are assigned to small groups, where students of about the save level or skill ability compete against each other. The games or activities of the small group may reinforce or teach new skills. After the game or activity is over, players return to their base team. Base team points are collected through the points earned during competition. Using this method of learning not only helps teach content, but also develops cooperative learning skills. This method of using games for learning appeals to those students who thrive on competition (generally male) and those who like cooperative learning (generally female).

Simulations games are popular in the classroom. A form of the simulation game, Real Life Math, is a favorite in many classrooms. It is the game of Life, adapted for classroom play. In this simulation game, students are sadly left orphaned on the day after high school graduation, with nothing but a bed, a TV, and $3,000.00. Using real newspapers, they must find a car they can afford, an apartment to share, and 3 places to apply for jobs. Their goal is to make the right decisions and earn enough money to survive in life. While boys love looking through the newspaper for great used car deals, girls enjoy apartment hunting. Many girls make up pets to take to their apartments. Often, as the days of the simulation game go by, girls begin to draw floor plans or start looking at advertisements for furniture. It begins to look very much like The Sims.

In Real Life Math, every student goes through three job interviews. People from the community come in and provide mock interviews for students. Overwhelming, female students dress up for the interviews and become very involved in role playing during the interviews. Fewer boys became caught up in the role playing.

Once students have a job, they open up checking accounts, pay deposits, and buy their car. After a couple of weeks, students receive bills and their first paycheck. They also draw life tickets that usually end up costing them some extra funds: late video fees, party expenses, etc. Trying to juggle their bills and their paychecks, students have a taste of real life with minimum wage jobs. Students learn many things during this unit. One comment made by a student was, "Now I know why my parents are so grumpy when they pay their bills." This is a fairly insightful observation for a fourth grader. The connections made when playing educational simulation games provide an excellent scaffold for learning.

In education, a simulation game can offer a player a chance to explore a concept or experience in an environment where variables can be manipulated and controlled. Science uses simulations for experiments (Gussin, 1995). Simulations can be used to explore history, large social concepts, or elements of literature. In order to be successful learning tools, educational simulations must capture the interest and engage the mind of learners. They are, however, very time consuming, and it is hard for a teacher to facilitate the direction and content of the intended and unintended learning that takes place. Games are much easier to manipulate for educational purposes. They provide learners with connections between experience and content.

How each gender uses games in the educational process can have an impact on the learning which will take place. Girls prefer inductive problem solving, discussion, and contemplation. Boys prefer visually stimulating environments when learning. On the whole, they like competition and beat the clock activities (Gurian, Henley, & Trueman, 2001). To be effective, a game should have elements that will appeal to both genders.

One of the games often used in classrooms is *Oregon Trail*. Both boys and girls find something of interest in its design. First, there is a storyline and characters that engage the imagination. Girls spend time carefully shopping for the trip west and reading the opening information while boys usually jump right in and began playing with minimal review. Once prepared, players travel west before bad weather comes or before they die. The first time one student played the game, he commented on how sad he was when he found out that his family was dying along the way. After that first game, though, he no longer cared if other characters died, as long as he survived and had money left at the end. As students play over several days, many girls keep their teachers informed about how their families were faring. Boys were quick to brag about the animals they shot or fish they caught. Both genders enjoyed this game and both learned about the hardships of traveling west during the Nineteenth century (Caftori, 1994). However, each came away with a different slant of the trip west. Overall, boys talked about the skills they would need: Shooting, fishing and being able to fix broken materials. In order to win, many boys felt that the early settlers didn't need education; real survivors needed money and skills. In contrast, while girls also talk about the skills that were needed, they often felt as if they had not really won if many of their family members died. They tended to focus on the deaths and the lack of medical knowledge that caused so many people to die. Educationally, however, it is hard to evaluate if the game actually teaches curriculum content. Players can spend large amounts of class time bartering, shooting and fishing, without ever learning geography or history (Caftori, 1994).

Another educational game that is well liked by students of both genders is *Gizmos and Gadgets*. In this game, students answer physical science questions and build a race car that competes in a race (Lindsey & Kaufman, 1998). The game requires students to choose the correct answer in a series of questions in order to get the parts to build the engine of the race car. Success comes in increments throughout the game. Girls seem to enjoy pitting their minds against the questions and receive satisfaction from solving the problems correctly, while boys often speed through the questions in order to get to the race. Both enjoy the game because it has layers built into it that are appealing. While both sexes gain knowledge, the boys often learn the material through trial and error. Most boys take a stab at an answer and if it works, they move forward. If the answer is wrong, they try again. Eventually they remember the correct answers in order to navigate their way through the game. Girls, on the other hand, spend a long time analyzing questions, talking it through, and even calling in the teacher, as a consultant. When they finished with the questions, the girls know the material, but they take longer to go through the game. These game strategies carry over in the ways boys and girls take tests. Multiple choice tests are much easier for boys than girls, because of their different styles when approaching problems (Gurian, Henley, & Trueman, 2001). Boys pick through the information and are more willing to take a chance by guessing, while girls prefer to know that their answer is correct before committing themselves.

4.8 Game Design for Education

While it is important to design games that appeal to genders traits at a basic, brain preference level, it is also necessary to analyze the desired outcomes of educational games. Because of its design, standardized testing appears to be slanted towards boys (Gurian, Henley, & Trueman, 2001). Instead of encouraging young women to play educational games that appeal to their strengths, it might be more important that the games they play help build those skills that are needed to develop strong standardized test skills. The purpose of educational games, as opposed to recreational gaming, is to improve learning. Thus, educational games require a blend of both male and female weaknesses, as well as skills. It is important in an educational setting to not only use educational games that reduce weaknesses but continue to build on strengths. If visual spatial ability is strong, but language skills need to be developed, a good educational design will incorporate both. A good learning tool provides positive feedback to students and allows them the opportunity to learn from their mistakes.

Designing a video game that will appeal to both genders has been difficult because it must appeal to both men and women at different levels. Video games are especially hard to design for females; since the software is not yet sophisticated enough to appeal to multi-tasking minds that traditionally excel in communication and contemplation. Boys are considered to be more auditory. When learning something new, girls prefer that the noise level of their study environment be very low (Honigsfeld & Dunn, 2003). Almost all video games are provided with sound tracks. Educational games will need to have mute buttons or removable headphones for those students who find it difficult to learn material with a sound track.

Most electronic games are designed to appeal to the visual spatial abilities of boys. They move very fast. A player must juggle many pieces of trivia that must be remembered and placed into a cohesive pattern. Boys prefer that the "whole screen changes at once, rather than in sections. They prefer to have a variety of choices on the screen and more control of the screen." (Passig & Levin, 2000, p. 67). Girls, on the other hand, were found to value colors on the screen and illustrations. Girls preferred red and yellow, while boys favored blue and green. Girls were not as interested in moving pictures, but in having many pictures on the screen. Girls also were more worried about using the computer appropriately, and valued the help button much more than boys (Passig & Levin, 2000). It is interesting to note that boys were much more interested in how to use the navigation buttons. While the buttons didn't matter to the girls, boys "preferred arrow buttons and square buttons. Girls like colored buttons" (Passig & Levin, 2000, p. 70), however, so an integration of the two might insure that both preferences are addressed. One company, Sony, has addressed this preference by designing play station controllers that include both icons and colored buttons.

The question of characters in educational games is tricky. Even when characters are designed to be gender neutral, children still "assign gender to the figures and the norm appears to be male." (Bradshaw, Clegg, & Trayhurn, 1995, para. 14). Characters within a game succeed when they confront opponents. Trying to provide characters that are free of any stereotyped bias may be difficult. By identifying with the characters within a game, children develop a stronger attachment to the action, which in turn helps to build connections to the curriculum within the game. Character identification can be a double edged sword. According to a study completed by Sadford and Madill (2006), male adolescents love video games because it allows them to resist authority in a socially acceptable manner.

When game designers put together their games, they probably don't sit down and ask themselves, "How can we provide an outlet for male adolescents to pit themselves against the established authority of their community?" Educators, however, must critically examine games in order to determine what learning will take place. Classroom teachers cannot afford to waste instructional time with games that are not helping students to learn. It is not enough for students to just play or win an educational game. The educational values within a game must be analyzed.

The diversity between boys and girls, as applied to learning, continues to be studied by neuroscientists and educators. In light of research into the neuroscience of both male and female brains, it is important that educators also look at the design and educational impact of games on boys and girls. When developing games, there are many things to consider in order to attend to the inclination of both genders. See Table 4.1 for differences in preferences. Finding games that appeal to both genders and promote learning is not easy. As more and more classrooms embrace games and technology, educators must examine them more critically than ever before. It is important to remember that every child has their own preferences and what appears to be gender bias in games and play is often blurred by societal expectations. There are more variations between individuals than between genders.

Table 4.1 Summation of Gender Differences Discussed

Male	Female
Strong Visual Spatial abilities	Strong Verbal abilities
Single focus	Multiple focus: stronger at Multi-tasking
Prefers to experiment rather than ask for help	Easy access to "Help"; hesitant to choose an answer/response if unsure
Prefers moving figures	Appreciates illustrations and colors on game screens
Prefer graphs, charts, & visual clues	Prefers text
Puts information into patterns	Puts information into words: connections with stories or people
Goal oriented	Enjoys Cooperative learning
Willing to experiment or guess; risk taker	Contemplative and intuitive
Prefers blue and green in games	Prefers red and yellow in games
Competitive	Less aggressive
Quick reflexes; eye hand coordination, response to stimuli	Strong fine motor skills
Prefers to look for patterns	Prefer concrete details
Prefers square and arrow buttons	Prefers colored control buttons
Prefers a variety of choices on game screens	More sensitive to feelings/emotions in social situations
Prefers to spread out when working	Prefers quiet environment for serious study
Prefers games that use strategy to win	Enjoy rhythmic, verbal games

4.9 References

Anderson, M. (2005). Sims 2 fans take spotlight. *AdWeek* 46(4), 34.
Bell, R.C. (1979). *Board and table games from many civilizations.* New York: Dover Publications.
Blatchford, P., Baines, E., & Pelligrini, A. (2003). The social context of school playground games: sex and ethnic differences, and changes over time after entry into junior school. *British Journal of Developmental Psychology*, 21, 481-505.
Bonanno, P. & Kommers, P.A.M. (2005), Gender differences and styles in the use of digital games. *Educational Psychology*, 25(1), 13-41.
Caftori, N. (1994) Educational effectiveness of computer software. *T.H.E. Journal*, 22(1), 62-66.
Canadian Teacher's Federation. (2003). *Kids take on media: Summary of finding.* Retrieved January 15, 2007 from www.ctf-ee.ca/en/projects/MERP/summaryfindingd.pdf
Elkonin, D.B. (1999). On the historical origins of role play. *Journal of Russian and East European Psychology,* pp. 49-89.
Erlauer-Myrah, L. (2006). Applying Brain-Friendly Instructional Practices. *School Administrator, 63*(11), 16-18.
Gee, J. P. (2003). *What video games have to teach us about learning and literacy.* New York: Palgrave Macmillan.

Grossman, L. & Song, S. (2002). Sims nation. *Times South Pacific*, 48(56). Retrieved June 3, 2007 from Ebsco.

Grugeon, E. (2005, April). Listening to learning outside the classroom: Student teachers study playground literacies. *Literacy*, 3-9.

Gurian, M., Henley, P. & Trueman, T. (2001). *Boys and Girls Learn Differently!* San Francisco, CA: Jossey-Bass.

Gurian, M., & Stevens, K. (2006). How Boys Learn. *Educational Horizons, 84*(2), 87-93.

Gussin, L. (1995). Constructive lessons: building and playing simulation games. *C. D. Rom Professional*, 8(5), 40-50.

Hayes, E. (2005). Women, video gaming, and learning: beyond stereotypes. *Tech Trends,* 49(5) 23-28.

Honigsfeld, A., & Dunn, R. (2006). Learning-Style Characteristics of Adult Learners. *The Delta Kappa Gamma Bulletin, 72*(2), 14-17, 31.

Honigsfeld, A. & Dunn, R. (2003). High school male and female learning-style similarities and differences in diverse nations. *The Journal of Educational Research*, 96(4), 195-206.

It'll never fly:Monopoly. (2007, May). *Management Today, 14.*

Jordan, E. (1995) Fighting boys and fantasy play: the construction of masculinity in the early years of schooling, *Gender and Education, 7*, 69–86.

Kuther, T.L. & McDonald, E. (2004). Early adolescents' experiences with, and views of, Barbie. *Adolescence*, 39(153).

Leach, B. (2007). Action figures: the history of Star Wars collectibles. http://actionfigures.about.com/od/actionfigureprofiles/a/starwars12.htm Retrieved December 29, 2007.

Lindsey, D. & Kaufman, J. H. (1998). Science. *Teacher Magazine,* 9(4).

McLean, C. (2006, October). Playtime: Learning IT through gaming. *Certification Magazine,* 8 (10), 60-29.

Moltenbray, K. (2006). Casual approach., *Computer Graphics World*, 29(4). Retrieved June 2006.

Moreno, J. D. (2006). The Role of Brain Research in National Defense. *The Chronicle of Higher Education*, 6-7.

Otnes, C., Kim, Y.C., & Kim, K. (1994). All I want for Christmas: an analysis of children's brand requests to Santa Claus. *Journal of Popular Culture*, 27(4), p. 183-194.

Orme, N. (2001) Child's play in medieval England. *History Today.* 51(10). pp. 49-55.

Park, B. (2006). The Science of Learning Meets the Art of Teaching. *Education Canada, 46*(4), 63-66.

Passig, D. & Levin, H. (2000). Gender preferences for multimedia interfaces. *Journal of Computer Assisted Learning*, 16, 64-71.

Pelligrini. A.D., Blatchford, B, Kato, K., & Baines, E. (2004). A short-term longitudinal study of children's playground games in primary school: implications for adjustment to school and social adjustment in the USA and the UK. Social Development, 13(1), 107-123.

Rheingold, H. & Cook, K. (1975). The contents of Boys' and Girls' rooms as an index of parents' behavior. *Child Development.* 46, p. 459-463.

Sanford, K. & Madill, L. (2006). Resistance through video games: It's a boy thing. *Canadian Journal of Education,* 29(1) 287-306.

Scanlon, J. Ed. (2000). *The gender and consumer culture reader.* New York University Press. NY.

The Sims. (2003, February). *Electronic Gaming Monthly*, 163, 140.

Sweedyk, E. & de Lael, M. (2005). Women, games, and women's games. *Phi Kappa Phi Forum* 85(2), 25-28

Thorne, B. (1993) *Gender play: girls and boys in school.* New Brunswick: Rutgers University Press.

Chapter 5
Video Game Pedagogy:

Good Games = Good Pedagogy

Katrin Becker

Abstract

We have always appropriated whatever technologies are available to us for use as technologies for instruction. This practice may well date back as far as human communication itself. The practice of "studying the masters" is also an old and respected one, and using this perspective we can take advantage of the opportunities afforded us in studying outstanding examples of commercial digital games as "educational" objects, even if they weren't produced by professional educators. By examining successful games through this lens we can progress towards an understanding of the essential elements of 'good' games and begin to discuss the implications this holds for the deliberate design of educational games. There is, however, a caveat: knowing why a game is good is not the same as knowing how to make a game good. It is nonetheless an essential step in that process.

This chapter examines some ways in which a few "good" games implement some well-known learning and instructional theories. "Good" games in this context are defined as those that have experienced both substantial commercial success and broad critical acclaim: the deliberate implementation of one or another learning or instructional design theory is not a prerequisite. In fact most will not have been consciously influenced by formal educational theory at all.

The implications of this study include the notion that learning and instructional design are compatible with good game design and vice versa. Finally, this chapter will present some key distinctions between digital games and other learning technologies and what this might mean for the development of design models and methodologies.

5.1 Introduction

Anyone who makes a distinction between games and learning doesn't know the first thing about either.

- Marshall McLuhan

That games have the potential to be effective learning technologies is no longer news. Such games are already out there: *Making History* (Muzzy Lane Software, 2006), *A Force More Powerful* (BreakAway Games Ltd., 2006), *Real Lives* (EducationalSimulations, 2002), *Big Brain Academy* (Nintendo Co. Ltd., 2006), and *Global Conflict: Palestine* (Serious Games Interactive, 2007) are just a few recent examples. That games deserve a place in the classrooms of our schools and institutions of higher learning is not quite as clear.

The goal of this chapter is to make that second connection a little stronger. Interest in games for learning in formal education is high but so is suspicion. Neither should be surprising. Talk of digital games for learning seems to be everywhere now. The Association for Educational Communications and Technology (AECT) featured several panels and sessions focusing on games in education at its annual conference in 2006 and will provide even more in 2007. Most organizations that deal with the use of technology in learning now feature articles on games and education and several publications have devoted entire special issues to gaming (Journal of Design Research v5(2) 2006, British Journal of Educational Technology (in press), ACM Journal of Educational Resources in Computing (upcoming), AECT Tech Trends, v49(5) 2005, Journal of Media Literacy v52 (1&2) 2005, to name just a few). Hardly a week goes by that one does not find an article on games in newsfeeds devoted to education and formal schooling. Large and highly respected organizations are saying we need to use games in school. The recent report published by the Federation of American Scientists stated: "There was strong consensus among the Summit participants that there are many features of digital games, including game design approaches and digital game technologies, which can be applied to address the increasing demand for high quality education" (Federation of American Scientists, 2006). At the same time, suspicion remains high: games are blamed for youth obesity (Reitman, 2003) while simultaneously offering promise as a means to combat the very malady for which they are blamed (Lash, 2006), and of course concerns over violence in games and their effects remain strong (Carolipio, 2006; Minton, 2006). With so many conflicting voices, it is important to reassure teachers that games *can* still adhere to the principles of good learning that they have been taught.

Although this is changing, most teams currently engaged in the creation of games for learning have key members who come from the games industry and so they bring to the table considerable expertise in commercial game development where success comes from pleasing the audience. Commercial game buying decisions are based heavily on game demos and word of mouth (Dobson, 2006) so in one way or another it is the game itself that determines its sales, and ultimately its survival. Game design is critical to game success. The typical shelf-life for commercial games is still on the order of 6 months (Kücklich, 2005) so most don't get second chances. As the interest in using games for learning becomes more common, the demand for new and specific games will grow, and with that demand for design teams able to create these games will also grow. The number of people from the games industry interested in creating games specifically for education is small so we will need to fill that gap by training educational games designers. Along with instructional games designers, there will also be a demand for ways to teach teachers how to become familiar with and make effective use of the medium. One step along that path is to connect the dots between what we already know and do in education and what is currently being done in game design. Educational games design must be a synergy of both game design practices and instructional design practices – neither can be layered on top of the other, and neither can be subordinate to the other either.

5.2 Studying the Masters, and the Scholars

"One of the most difficult tasks men can perform, however much others may despise it, is the invention of good games. And it cannot be done by men out of touch with their instinctive selves."

- Carl Gustav Jung[1]

If we were to take a close look at how the different forms of both classical and modern communication media (theatre, literature, film, television, etc.) have been used for educative purposes and which 'commercial' examples have been appropriated by educators, it becomes clear that the majority of the most remarkable and effective "lessons" taught to us in this way have been created by extraordinarily talented writers, playwrights, directors, and producers together with their teams (Hemmingway, Twain, Spielberg, Dickens, etc.). One other notion stands out. These significant educational works have, by and large, not been created by professional educators or instructional designers. What does this mean? Should we ignore what instructional design methods and theories have to say? The answer is, "Of course not". Far from trying to circumvent what educators and instructional designers have learned, we should recognize the opportunities afforded us in studying these outstanding examples as "educational" objects, even if they weren't produced by professional educators. We should try to characterize what it is about them that make them have the impact they do.

The case for studying the masters doesn't really need to be made, but let's review anyways. Looking at the practice of the 'masters' is an accepted approach to

---

[1] as quoted by Laurens van der Post in Jung and the Story of Our Time (New York: Vintage Books, 1977), pp. 411.

education in Fine Arts, the Performing Arts, Literature, and Music as well as a few others. All have a long tradition of learning from the masters. Although the reasons are rarely articulated, we often do the same thing when learning about leadership and Architecture, and one of the most significant modes of learning in Law is still to study famous and ground-breaking cases. They are studying their masters too. When it comes to the Sciences, although we do pay some homage to our icons and eponyms, we don't treat them as guides for the practice of Science – we focus on the results of their research.

In Education, we study the scholars and eagerly try to assimilate their theories. But we tend to align ourselves more with Science than with Art and tend not to study teaching by looking at how the best teachers do what they do. We typically try to adopt a more "scientific approach", especially in educational technology. As a medium, games are more closely aligned with film and theatre than they are with the more traditional learning technologies like textbooks or even websites. Given that, one could argue that much can be learned about how to design games by looking at the masters of this profession, namely the best games. By "studying the masters", we can progress towards understanding the essential elements of 'good' games and begin to discuss the implications this holds for the deliberate design of educational games. There is, however, a caveat: knowing why a game is good is not the same as knowing how to make a game good. It is nonetheless an essential step in that process. This chapter will make that step.

5.3 Connecting the Dots

One learns by doing a thing; for though you think you know it, you have no certainty until you try.

- Sophocles

Previous work by this author (Becker, 2006b) has focused on connecting commercial video games to accepted pedagogy in a fairly general way. It is not especially difficult to cherry pick specific elements from a wide variety of games in order to support an argument, and while this is useful it also has its limitations. Suppose we are going to examine Gagné's (1985) Nine Events . If we show that nine different games each implement one aspect well, we have still not shown that any one game is capable of embodying Gagné's theory. Further we have no evidence that ANY game that was able to incorporate all nine events would still be a popular game. We all remember films for example that may have had one or two good moments but are otherwise unremarkable or even bad. It is a much more significant feat to look at the whole of a work and see how the various parts fit together.

These steps towards establishing the pedagogy of games are important because the last time we tried to use digital games in education it didn't go so well and a part of that reason was that we did not understand the medium of the digital game (Egenfeldt-Nielsen, 2005). Games are far too complex to create to end up having them ignored because the target audience finds them tedious, boring, too difficult or too easy. To hear the kids talk, we already have enough learning objects like that. We probably don't need any more. What readers can take away from this chapter is evidence that good games *already* implement sound pedagogy, and on some level, this implementation of sound pedagogy is in fact a major contributor to what makes it a successful game. It is comforting to know that the implementation of sound pedagogy can lead to a compelling game – this may mean that this is also true of other learning technologies. That's not to say that educational games should strive to become commercially successful like their pure entertainment cousins – very few educational objects receive great commercial success. We are working with an entirely different scale than the video game industry at large. They make games that have development teams numbering in the dozens and budgets in the millions of dollars. They also expect profit in the millions. On the other hand commercial success and educational efficacy are not necessarily mutually exclusive.

This chapter examines several commercially successful games using instructional design theories and models as a lens. The implications of such an examination are two-fold:

1. Games work as instructional technology. Thus, established learning and instructional theory can be connected with current design practices in this new medium. Since claims are being made that digital games should be viewed as viable technology for use in education, this forms one facet of the necessary proof of concept.
2. Instructional design works for games. We can verify that games designed along learning and instructional theory lines can and do result in artifacts that remain compelling as games. This does not mean that ID methods can be followed like recipes to produce successful games. We have not yet discovered a formula for generating blockbuster movies or classic literature either, but we still value formal training in film-making and in writing. It helps develop better writers, playwrights, and film makers.

5.4 On Choosing Games for Study

We need to consider whether we are educating children for their futures or our pasts.

Geoff Southworth 2002

Why is it important to justify the choice of game being used as an example in a scholarly article or for the purposes of study? In the early days of games studies there seemed little call for careful scrutiny of one's game choices. We studied what we had handy and wrote about the games we were already playing. However, if we want to make the case that the game in question is *good* on some level (however we decide to define "good") then we really should have some evidence to back this up. If we are proposing the use of a game in the classroom or the study of some specific game to learn something applicable to our agenda, then we have a responsibility to explain why *that* game is suitable for our purpose.

Critical and commercial success are both recognizable and accepted (albeit subjective) measures of a game's popularity, and popularity in turn gives some indication of that game's perceived quality as judged by players, developers, and game critics. Combining a number of different measures to come up with a single measure is one way to ensure that games that end up at the top of the final list qualify as successful by more than one measure and have been assessed by more than one source. Since the main premise for examining commercial games in this study is to look at how games teach and otherwise support learning by studying 'at the feet of the masters', there must be some way to convincingly determine that the games from which the final choices are made are of a stature that would qualify them as among the best. This is not a straight forward task. In a sport like sprinting, determining who the fastest sprinter is can be done quite objectively – it is a matter of comparing competition times and the runner with the fastest time wins. No such objective measure exists for most creative endeavors, and since games are creative designs we can only produce subjective measures. Combining subjective assessments from many sources in a systematic way results in a list with which most (industry, gamers, critics) would agree.

In a recent article offering suggestions for how the Academy could build stronger ties with the Games Industry, John Hopson argues that we should "use examples from bestsellers. A good example from a popular game is more effective than a great example from something they've never heard of. Industry people often suffer from an "if-they're-so-smart-, why-ain't-they-rich" attitude towards smaller titles. Even if the small title is a perfect example of how the theory works, they're going to be less likely to listen if they haven't heard of the game ahead of time. Commercial success is one way of making sure that the audience will respect your examples, but you can also use titles that are well known or critically acclaimed but which weren't necessarily huge blockbusters. It's also important to keep your examples as current as possible, because many industry folks will see a three-year-old example as ancient history" (Hopson, 2006).

5.4.1 The Chosen Ones

Three games have been chosen for this study, all of which are well-known to gamers. These games were chosen for their style, and for their critical and commercial success. None are claimed to be of interest as educational games. Two of them, Animal Crossing Wild World (ACWW) (Nintendo, 2005) and Super Mario Bros. (SMB) (Miyamoto, 1985, 1988, 2006) consistently appear on "Best Games of All Time" lists. The third game chosen, Phoenix Wright: Ace Attorney (PW) (Capcom Co. Ltd., 2005) has appeared on fewer lists but in its defense it is also a relatively new game that does not have any predecessors to fall back on like the other two. In all the reviews that are available, Phoenix Wright has been consistently rated 80/100 or higher. Two of the games (ACWW, SMB) are rated 'E' for everyone, and the third (PW) 'T' for teen. It seems reasonable to assume that there is limit to what can be gained from studying games rated 'mature' when we are trying to understand how to make games for younger audiences, so no 'M' games were considered for this study.

Although all three are produced for the same device: the Nintendo DS portable console) they all belong to different genres with different styles of gameplay. Both *Animal Crossing Wild World* and *Phoenix Wright Ace Attorney* have a considerable amount of written dialogue even though neither includes any actual talking. *Phoenix Wright* uses several phrases, like "Hold it!", and "Objection!", but his can not really be described as talk. *Mario* similarly has very little talk other than a very few phrases, but unlike the other two it is almost completely devoid of text.

Throughout the following examinations, dialogue quoted from either of these games will be displayed indented quotation text, identified by 'speaker', and located within the game as far as possible. Formal citations of quotes are not possible in most digital games as transcripts are not published, so this will have to be sufficient. Every effort was made to be as accurate as possible, but since dialog was transcribed while the game was running, some minor omissions or inaccuracies are possible, and for that the author apologizes in advance.

5.4.1.1 Animal Crossing: Wild World

Animal Crossing: Wild World (Nintendo, 2005) is an open-ended RPG (role-playing game). The original version of the game, Animal Crossing (Nintendo, 2001) appeals to a wide range of ages both male and female and is on many Best Games lists. Animal Crossing: Wild World was released for the DS console, but is essentially the same as the original, with respect to the main story, controls, interface, and how it teaches players what they need to know. Ultimately ACWW was chosen for study over the original version for the convenience afforded by the small portable device as opposed to the standard console which must be connected to a television.

This game is situated in a small fictional village whose landscape includes trees, flowers, rocks, an ocean front, and a river with several ponds and waterfalls. Players may choose the name and gender of their character (avatar) as well as the name of the town in which the game takes place. The town has various locations where things can happen: a general store, a town hall, a clothing shop, a museum, and a main gate. Although gameplay is the same, there are some differences between the console and the DS version. For example in the console version there is also a town dump whereas the DS version has a recycling bin inside the town hall.

These are functionally equivalent as they behave much the same way and serve the same purpose. There are other townspeople besides the player which include various permanent residents as well as residents that move in and out from time to time. Curiously, the characters in this game take their houses with them when they leave! There are no required tasks or goals to achieve and no definitive end to the game; instead players decide for themselves which goals they want to pursue. When the game first begins, the player lives in a small house that has a small mortgage held by the owner of the general store. The local currency is called the "bell". All role-playing games and many other games have some sort of in game economy and this one is no different. Players can earn bells in various ways and can use that money to pay off the mortgage, which is held by the shopkeeper. Each time a mortgage is paid off, the player gets a new addition to her house along with a new, larger mortgage. Although players have no choice about the additions to their house, they can decide not to pay off the mortgage and pursue other goals instead, and they are not penalized for choosing this path. Other goals include various collections (fish, insects, fossils, pictures, clothing, furniture, and a few other items), cultivating flowers, growing trees, designing star constellations, or cultivating friendships among the residents. *Animal Crossing* is a game space where players are rewarded for tending.

5.4.1.2 Phoenix Wright: Ace Attorney

Phoenix Wright: Ace Attorney (Capcom Co. Ltd., 2005) has received consistently high ratings though is a relatively new title. Its format is different from most popular games today and that design makes it of interest for study. For one thing, it is surprising to many that a game featuring a lawyer as the main character should become so popular. Another is that this game is essentially a branching story that ultimately has only a single path to the end. This is a format that is potentially quite useful in education – gameplay is essentially 'on rails' with few opportunities for exploration outside of the main goals and yet this game remains very highly rated. Many of the criticisms of this game have to do with the low replay value afforded by the format, but while players expect to be able to play commercial games more than once, a lack of replay-ability may not be a detractor in an educational context - learners in a classroom situation may never have an opportunity to play the game again after they have completed it once. Another aspect of interest to educational game design is that it has very minimal animation so a game such as this could potentially be produced on a fairly low budget. This game is currently only available for the Nintendo DS portable platform.

In this game we play the role of Phoenix Wright, a newly-minted defense lawyer taking on his first cases. The game consists of five separate cases each involving a murder investigation that culminates in a trial which we must win. Each of the main characters has interconnecting back stories, bits of which are revealed from time to time through the five cases. In all cases the person accused of the crime is innocent and it is our job to gather evidence and other clues which will be used to argue our case during the trial in order to have our client found not guilty. We must also discover who the real killer is. It should be noted that although it is possible to learn some of the terminology associated with the legal system through playing this game, the game's designers make no claim as to the accuracy of the court procedures or any other legal aspect of the game.

5.4.1.3 The New Super Mario Bros.

The New Super Mario Bros. (Miyamoto, 2006) is a 'simple' platform game whose original version is more than 20 years old. This game is also on many top 100 lists. The particular version used for study is a remake of the original *Super Mario Bros.* (Miyamoto, 1985). Mario is a game for which any claims to educational value could only be made by either: 1) a profound exaggeration of the content of the game or 2) a radical expansion of the definition of education. Nonetheless, this game is still of interest in the current context for a number of reasons. First, it has remained consistently popular throughout its various incantations since it was first released over 20 years ago and that kind of staying power deserves further examination. Secondly it bears a strong similarity of form to one of the most popular educational games for that same time period, namely *Math Blaster* (Davidson, 1986) and while *Super Mario Bros.* continues to garner praise from game designers and gamers alike, *Math Blaster* is in many ways its antithesis (Becker, 2006a). Why is one so popular with almost everyone while the other is popular among teachers but panned by everyone else? A closer examination of Super Mario Bros. and its sequels may reveal some clues.

This game has a very simple premise, Bowser and Bowser Jr., the bad guys, have kidnapped Princess Peach, and it falls to Mario to rescue her. Virtually all of the Mario games are two-dimensional platform games where Mario can move from side to side and sometimes up and down but all movements and action occur in a flat, two-dimensional space. The nature of the game's challenges could perhaps best be described as those of an obstacle course. There is an obvious beginning to each course, obstacles (environmental and other characters and objects) to be avoided or neutralized (no-one dies in this game) and an end goal to reach, all within a pre-determined time period. As in a real life obstacle course, the obstacles themselves may have no logical connection to the ultimate goal – they are merely things that are trying to prevent us from reaching the end. *The New Super Mario Bros.* contains 80 different courses (levels) spread across eight different "worlds", where the set of courses in one world share similar landscapes and challenges. When the game begins, we see Bowser run off with Princess Peach, but then we do not see her again until we reach the end of the courses in the first world.

5.5 Game Elements

If you must play, decide on three things at the start: the rules of the game, the stakes, and the quitting time.

Chinese Proverb

One of the challenges in examining commercial games as learning objects is that nether structure maps conveniently onto the other. Game design and instructional design both come in many shapes and sizes, but both forms have more structures and elements in common with other designs of the same field that they do with designs of the other field. Thus most games share many structural elements that serve various purposes within those games and looking at existing commercial games as instructional technologies requires an understanding of those game design elements so that their connections to instructional design can be identified. In aid of that the following terminology (see Table 5.1) is included as reference and to clarify how the terms are being used in the remainder of this chapter. This is by no means a complete list of game elements and only terms that relate directly to the type of examination being conducted in this chapter are explained.

Table 5.1 Brief Glossary of Structural Game Elements

GAME ELEMENT	DESCRIPTION
A.I. Artificial Intelligence	The core 'engine' of the game that embodies the game's rules and conditions for winning, as well as how the characters within the game will interact with each other.
Attract Mode	This mode is the one that runs when the game is on but not in a state of active play. In arcades, this mode is the one always running when no-one is playing it. Some console games also have an attract mode that runs until the player restarts or continues the game.
Back Story	The story that underlies the game, and sets the stage for the main game goals.
Boss Challenges	These are challenges (often physical conflicts) with a major opponent and often mark the final challenge of a level or the entire game. Many games require players to achieve a certain level of achievement or score in order to earn the opportunity to enter a boss challenge.
Cut Scenes	These are non-playable parts of the game where part of the back-story or game narrative is revealed, typically in small portions lasting anywhere from several seconds to a few minutes. They can be in the same style and quality as the game itself, but they can also appear as movie quality clips.
Game Rules	These are the fundamental mechanics and dynamics of the game and its behavior.

GAME ELEMENT	DESCRIPTION
H.U.D. Heads Up Display.	Commonly used to refer to the display board that contains the game's vital information such as score, the player's statistics (health, assets, etc.), current game conditions, and so on. This may also include a map and other information.
L.O.D. / P.O.V. Level of Detail / Point of View.	Games typically allow players to change the level of detail by zooming in or out. It may also be possible to chage the point of view so players can see what is behind them or look at objects from a different angle.
Levels	Somewhat similar to chapters in a book, levels are parts of a game that contain one or more complete challenges. Subsequent levels typically build upon previous ones by adding new or more difficult challenges, new abilities, opening up new areas to explore or adding new opponents. Level progression goes from simple to complex or easy to hard.
N.P.C. Non-Playable Character	A character that appears in the game with which you may or may not be able to interact but whose behaviour is determined by the game's design. These characters are not controlled by the player.
Narrative	The ongoing story as it does or can unfold. It is what comes after the back-story, often adding to it.
Outcome	The outcome is the final state of the game. This is a quantifiable (i.e. obvious) state: it will be clear whether or not the player achieved the stated goal. (Salen & Zimmerman, 2004, p. 96). The win state always depends on the valorization of the game. MMOs and other persistent-world games tend not to have an end outcome, but will almost always have missions, quests, or mini-games that do have clear and definite outcomes. (Juul, 2005)
Perspective	First-Person (player as character); Third-Person ("over-the-shoulder"); Top-Down (bird's-eye view); Isometric (tilted top-view; slightly to the side) ; Side-View (two-dimensional horizontal view)
Sandbox Mode	Practice mode, where scores do not count towards a win. Some games contain ony a sandbox mode as their primary mode of gameplay, such as the SIMs games.
Story Mode	That part of the game where gameplay is "on-rails", meaning that the player has little to no control over where they go and what tasks they attempt. They are given specific tasks which must be completed, often to a pre-determined minimum level of competence in order to progress. This device is often used to ensure that the player is exposed to specific story elements, and often makes use of cut-scenes.
Time: actual and game-time	The passage of time in games may change between actual real-world time and accelerated, skipped, or even slowed game-time. Often the passage of time during play is reflective of real time, but like in movies, a change of scene or location can also coincide with a change of game time.

GAME ELEMENT	DESCRIPTION
Trailers	These are the game advertisements, often containing cinematic quality clips, screenshots of actual gameplay, and other dramatic devices to give potential players an idea of what the game is like.
Tutorial Mode	Often occurring at the beginning of the game but in some games it can also be triggered at the start of a new level or challenge or in response to poor player performance. In this mode often the player often receives direct guidance, visual, verbal, and otherwise from the game. This mode's purpose is to help the player acquire sufficient knowledge and skill to mange the basic gameplay.
Valorization	Different values are assigned to different outcomes within the game; some are winning outcomes (better) and some are loosing outcomes (worse). Often the more highly valued outcomes are more difficult to achieve than the negatively valued outcomes. (Juul, 2005) The values placed on various outcomes as well as the values associated with various choices made during gameplay are determined by the game designers, and may or may not coincide with societal norms, or the value-set personally espoused by the designer.

5.6 Learning and Instructional Design Theories and Models

The heart of this chapter is broken into two sections: one that deals with several older, well-known, and relatively structured learning and ID models and theorists. Their structured nature allows for relatively detailed analysis both from a generic game perspective and from a more detailed game-case study perspective. Each of these will be examined in both ways. Each of the first three case studies includes a visualization that depicts how these game elements connect generally with elements of the ID theory/model in question as well as a description of how the model can be mapped on to a specific game. The second section examines several theories that are somewhat less linear and for these there is no general visualization. Instead we will go straight to the case studies after a brief explanation of the main tenets of the theory.

In an analysis such as the one that follows in this chapter, there is a necessary trade-off between describing the detailed analysis of one model and one game and offering a more cursory comparison using several examples of each. The second approach is the one to be used here, as it was felt that a few games able to stand up to scrutiny from multiple angles will provide stronger support for the main premise of this chapter, that being that good commercial games already demonstrate the properties of sound instructional design, even in games not intended as educational entities. Throughout the following pages, these connections will be tied together with explicit examples from the games to illustrate the connections.

5.7 The Classics Revisited

There is no subject so old that something new cannot be said about it.

- Fyodor Mikhailovich Dostoyevsky

There are three models used to shape our study of three specific games in the first section. These three models have been used for this purpose before (Becker, 2005, 2006b) but the examples from games were used in a very general sense. Each of the chosen games is paired up with one of the chosen models to show how 'they got it right'. Many games employ various strategies and there are often multiple support mechanisms for each strategy.

5.7.1 Gagné's Nine Events of Instruction

The work of Robert Mills Gagné (1916-2002) hardly needs any introduction; next to Bloom's Taxonomy (Bloom, 1964) it may well be the best known instructional model in existence for teachers in North America. Gagné's nine events of instruction (Gagné, Briggs, & Wager, 1992) still functions as a useful guide for the design of instruction and it is perhaps fitting that this be the first instructional design model used in our examination of the masters. Game elements can directly and indirectly embody all elements of this model, as is indicated in Figure 5.1. Connecting game elements to instructional design model components serves to illustrate which game elements can be shown to implement accepted instructional approaches, and as noted, this work has been described by this author elsewhere (Becker, 2006b). However, while the previous work did use examples from many different games, it did not analyze any single game to see how it measured up. The next section takes a more detailed look at Gagné's nine events and how one specific game fits.

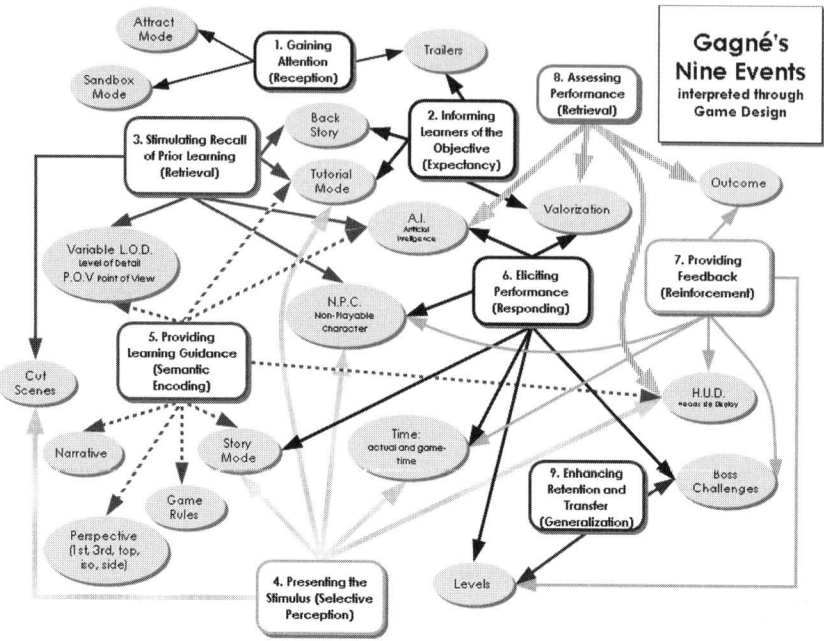

Fig. 5.1 Connections between Gagné's Nine Events and Common Game Elements

5.7.1.1 Gagné's Nine Events of Instruction as expressed in Phoenix Wright

5.7.1.1.1 Gaining Attention: Reception

Capture their attention. The process of gaining attention in games often begins long before the game is even released. Trailers and demos are important for providing advance knowledge of many aspects of a game including the style, back story and main objectives. Just like movie trailers are designed to entice people to watch the film in the theater, game trailers are designed to entice people to buy. However, game advertisement as well as game reviews have only limited influence over game sales - word of mouth plays a significant role and for that the experience of playing the game must be good enough for players to encourage others to buy (Dobson, 2006). Thus, the opening sequences of any game must keep the players attention while providing sufficient guidance to avoid too much frustration.

In the event that players have not seen all the trailers or read the reviews, the game opens with a cinematic device that almost always grabs attention. Phoenix

Wright (Capcom Co. Ltd., 2005) opens with a cinematic clip of sorts - there is very limited animation and the artistic style and quality is identical to that of the rest of the game. The style is similar to that of anime - somewhat cartoon-like. The first episode is called, "The First Turnabout" and opens with images of a crime scene.

> *gasp*... *gasp*...
> *[A statue dripping with blood and a woman's body lies in a widening pool of blood set the scene.]*
> Dammit!...why me?
> *[A man stands over the body holding the statue. We see his face.]*
> I can't get caught... not like this!
> I-I've gotta find someone to pin this on!
> Someone like.... HIM.
> *[We see a silhouette of another man in the hallway.]*
> I'll make it look like HE did it!

The next screen displays text that gives the date: August 3, 9:47 AM; and a location: District Court Defendant Lobby No. 2. As with most games screen advances during dialog sequences are largely under player control. A small amount of text is displayed and then the player chooses to advance to the next screen when ready. If the game designers have done their jobs right, the player is intrigued by the mystery and continues the game.

The mechanism used to gain attention in this game is very similar to that used in film and television, albeit on a much lower budget. Given the murder-mystery lawyer-show approach to the story, the introduction is consistent with the style. Consistency of style is important if we are to ensure that players buy in to the game and accept the premise.

5.7.1.1.2 Informing Learners of the Objective

Let them know what is expected of them. We already know that our overall objective is to investigate the case and free our client. We learned that previously from either the trailers, or through word of mouth, or both. What we still need to know is the details of the case. The next scene shows the lobby and a double door being guarded by what appear to be security guards. An exchange of dialogue follows which introduces Phoenix, Mia Fey: his boss, and Larry Butz: the defendant.

The opening dialogue sequences serve to introduce the first few characters and to give us some hints about their character and the relationships between them. We find out that Phoenix has a boss named Mia Fey, this is his first trial, and that he and the defendant share some history for which Phoenix feels he owes his client a debt of gratitude. The way this is presented visually is that the background scene remains quite static, and the character we are to meet (either Mia or Larry) appears 'center stage' for a time while the dialogue continues. The person we see does not change with the dialogue so we cannot use that to tell who the speaker is; instead each dialogue 'bubble' is labeled with the speaker's name. In this game, the animation serves more as punctuation for the dialogue than anything else. The game is very text heavy.

> *[Scene ends; screen goes black]*
> Phoenix:
>> My name is Phoenix Wright.
>> Here's the story:
>> My first case is a fairly simple one. A young woman was killed in her apartment. The guy they arrested was the unlucky sap dating her: Larry Butz... my best friend since grade school. Our school had a saying: *"When something smells, it's usually the Butz."* In the 23 years I've known him, it's usually been true. He has a knack for getting himself into trouble. One thing I can say though: it's usually not his fault. He just has terrible luck. But I know better than anyone, that he's a good guy at heart. That and I owe him one. Which is why I took the case... to clear his name. And that's just what I'm going to do!
>>
>> Now we know.

5.7.1.1.3 Stimulating Recall of Prior Learning

Remind them of what they should already know. In the first case of the game no assumptions are made about prior learning beyond the basic device controls. This game does not have any other games like it, so at the start of this case we are essentially starting fresh. When we get into subsequent cases we will get reminders of things we can do, but that will be described retention and transfer. At the start of the game we are essentially 'fed' the background information we need to know to begin this case. This is done through the dialogue exchanges between the first three characters that were introduced.

5.7.1.1.4 Presenting the Stimulus

Present the content. The way content is presented in this game and the kinds of visual queues that accompany it are quite differently presented in this game from most others. It has already been mentioned that the animation is minimal. The format resembles a comic book more than a dynamic game. Most screens display what amounts to talking heads: the character that is currently the focus of attention stands in front of a relatively static background image. It is not until we get a little further into the game that we find out that in many cases the background image (the 'scene') can be examined more closely. This is one way that players can interact with the game. When we choose to examine a location, the image of that location is loaded into the bottom screen of the DS (this is the one that is touch sensitive) while the character in the foreground remains in the upper screen (with the background still visible there too.) The image on the bottom screen then behaves somewhat like a webpage: there are certain 'hot spots' on the screen where players can examine the scene more closely or read more information about some item or visible object. We may click on a desk for example and be told about some event that occurred there earlier, or we may discover that there is some object inside the desk which then becomes an item of evidence we can take away with us when we leave the scene.

5.7.1.1.5 Providing Learning Guidance

Help learners encode and assimilate what they have learned. Our tools for this game are kept in the 'Court Record' which serves the same purpose as the inventory of other games. Most games provide some place where players can store items they acquire during the game. The Court Record is divided into two sections: one to hold items we find during our investigation, some of which we will later be guided to present in court or to other witnesses. The other section keeps information on the people we have met during the game. At the start of the game our Evidence consists of Phoenix's attorney badge and Cindy's autopsy report. We can click on either one to get more information about the item. The autopsy report gives the time and cause of death, but little else. The Profiles section contains information on three people: Mia Fey (our Boss), Larry Butz (the defendant), and Cindy Stone (the victim). An important aspect of this game as in any mystery is the character development and all the people we encounter in this game have peculiarities.

As is typical of many games, certain options are available only at certain times. During a trial sequence when the witness is giving a statement, we can not interrupt or do anything other than page through the dialogue screens. However, when it is our turn to cross examine the witness, we can either press for more information or present some piece of evidence to point out a contradiction after each statement. Guidance is also offered within the context of the game's story as well: if we seem to be getting off the track or missing some important connection, Mia appears to offer advice.

5.7.1.1.6 Eliciting Performance

Make them practice what they have learned. The first case takes place entirely in the courtroom. As with most games, the first level is simple and takes relatively little time to complete. It serves as orientation to help players understand the game's interface and acquire the basic skills they will need to play the game effectively. The gameplay choices and 'courtroom procedure' as they exist in Phoenix Wright's world are shown and practiced in the first case. The judge (who is the same for all five trials) gives us a short tutorial to help us figure it out, all embedded as part of the story. The judge comments that we (Phoenix) look nervous so he will give us a short test. We are asked three questions, the answers to which can be located among the information we already have in our Court Record. One question we are asked is the name of the victim – there is a profile on her in the court record to which we have access. If we answer the question right, we go on to the next question, but if we answer the question wrong, our boss, Mia Fey tells us how to find the correct answer, all in character of course:

> *Mia*
> I think I feel a migraine coming on. Look, the defendant's name is listed in the *Court Record*. Just touch the *Court Record* button to check it anytime, okay? Remember to check it often. Do it for me, please. I'm begging you.

The tutorial is not repeated after the first case. As the case progresses, and especially in later cases, the mysteries that must be solved and the details that must be remembered become quite complex and substantial – players must be aware of clues and remember inconsistencies in evidence and testimony.

5.7.1.1.7 Providing Feedback

Tell them how they are doing. There is of course music but it is not strictly needed to play this game and the author played through most of the cases with the sound turn off. This did not noticeably detract from the game, although the sound was amusing. Through the short tutorial, we are introduced to the controls of the game and the significance of some of the details of the interface, such as red colored text, which indicates a clue or important evidence, and that Mia appears to give us hints if we do something wrong. Sometimes the judge also tells us when we have answered incorrectly, and we discover that we cannot proceed in the game until we give the correct response. Later in the game we also discover that we will not be allowed to proceed to the next chapter until we have gathered all the evidence we need and talked to all the witnesses we have to hear from. This game is far more structured and offers the user far fewer choices than most modern games do, and yet this game has become one of the highest rated titles of its year. Clearly, the format in and of itself is not a design liability.

5.7.1.1.8 Assessing Performance

Let them know how they did at the end. This is a game 'on rails', which means we must come to the right conclusion and there is only one path to winning this game. In fact there are only two possible outcomes: a guilty or not guilty verdict and if we end up with a not guilty verdict we will be made to repeat what we've done until we get it right. We will also be made to repeat individual cross examination sequences until we either find the right evidence to present, or run out of chances. We can repeat the sequence as often as we like so long as we do not make any unfounded objections. We have a limited number of chances to win each section (usually five – displayed as exclamation marks on the screen) and each time we present evidence that does not help us or raise an objection that is faulty we loose one of those chances. When we run out of chances, our client is pronounced guilty and we must start again from the beginning of the last chapter. We get frequent and immediate feedback on all of our actions, and each time we loose a round, we are given the option of starting back at the beginning of the chapter or the last place where we saved the game. We can save the game at any point where we can normally interact in someway, so we can set a save point almost anywhere.

As is common in most games, we may lose, but we will never be prevented from trying again. Part of what encourages such persistence in gamers is that we always know there IS a way to win, and that we can keep trying until we get it.

5.7.1.1.9 Enhancing Retention and Transfer

Help them remember and apply what they have learned. There are several levels at which retention and transfer occurs in games. The functional details of how to operate this particular game are useful primarily for the duration of this game and any sequels that might follow. More generally speaking, game genres contain similar functional interfaces as well as similar goals, challenges and reward structures. Just as learning about the structure of math textbooks in a general way helps us to get to the 'meat' of the next new math book we must use quickly, learning about the structure of various game genres allows players to get to the interesting parts of the next game faster. Having experienced *Phoenix Wright*, players are likely to be able to approach the next branching-story style game with certain expectations of what they can do and what they must do as well as what they should not expect. Ultimately this kind of generalization becomes habituated.

Finally, even though this game makes no claims about value beyond entertainment, there are still higher-order thinking skills that are being practiced in the course of this game, such as paying attention to and remembering detail; looking for contradictions in facts; and matching clues with events. While this game was not designed to be educational, the format, style, and even the lawyer premise could be useful devices to foster the development of these skills more realistically and accurately if used in an educational game than happens now in *Phoenix Wright*.

5.7.2 Reigeluth's Elaboration Theory

Reigeluth's elaboration theory (Reigeluth & Stein, 1983) is a macro level prescriptive strategy that builds on the work of Gagné (1977), Ausubel, Hanesian, and Novak (1978), Bruner (1966), Merrill, Li, and Jones (1991a) and many others of the late 60's and 70's. The goal was to integrate the then current knowledge on how to organize instruction in the cognitive domain. Many of the concepts unified in this model, such as the importance of selection and sequencing, instruction that progresses from simple to complex, and review strategies remain as relevant to modern teaching and instruction as ever. When viewed through this lens, digital games have many elements that connect with an elaborative approach to learning in many games, each level builds on the previous one, incorporating and building upon acquired skills and experiences. To show how this might be, the following concept map in Figure 5.2 shows each of the elements of the elaboration theory connected with various game elements. For example, learner control over both content and strategy is common in and is often embodied in the level of detail and perspectives available within a game, the player's ability to choose to focus on various aspects of a game and ignore others, and the ability to move around the game space. If we examine a single game we are likely to find the one game does not use all the available mechanisms to embody these elaborations, but in any good game, it is highly like that each elaboration will be supported in some manner. An examination of Animal Crossing (Nintendo, 2005) follows.

5.7.2.1 Reigeluth's Elaborations of the Animal Crossing Space

5.7.2.1.1 Organized Course Structure

Organize the 'course' structure to emphasize its primary focus. The primary underlying principle in this approach is the elaborative sequence from simple to complex, general to detailed, and abstract to concrete. For each distinct (single) type of content there will be a clear emphasis on concepts, principles, or procedures, and this can be seen in ACWW also. The main emphasis in ACWW is on procedures – doing things, and this is reinforced by the game's rule structure and its narrative as indicated on the back of the game box: "Whether you want to decorate your home, join in on special events, or just chat with the locals, there's always plenty to do!".

However, once players are familiar with the basic procedures, they can choose to focus on either procedures or principles, as one way to take up this game is to try and develop the relationships among the residents, which requires players to develop various theories and principles about how best to foster 'friendships' with the NPCs of the game. The game facilitates both approaches (procedural and principal), but the remainder of this example will focus primarily on the procedural aspects.

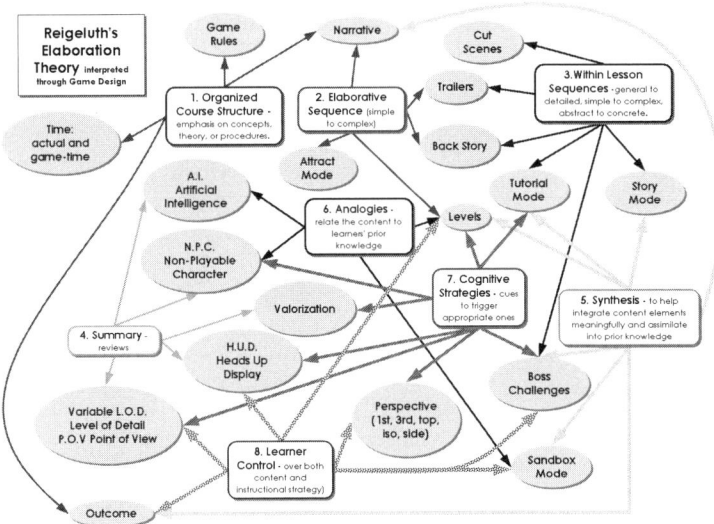

Fig. 5.2 Connections between Reigeluth's Elaboration Theory and Common Game Elements

5.7.2.1.2 Elaborative Sequence

Start with simple and basic ideas and progress to the more complex ones. The simple to complex elaborative sequence is presented initially through the game's back story, such as it is, which is presented to us at the start of the game. Most of the activities one can engage in are first introduced in a simple form. For example, the player's house begins as a small cabin. It is possible to decorate and furnish the main room of the house, but there is a limit on how many items can be placed in the house. As the player earns money and pays off the mortgage, the size of the house grows, and with it the number of items that can be placed inside. Further expansions add new rooms, which mean that each room can be decorated differently. At the beginning of the game we are given a small number of specific tasks to complete over which we have virtually no choice - we cannot continue with other activates until these are done. Our limited options are enforced by the fact that we have no tools or money and the fact that Tom the shopkeeper who hires us to work for him is also the character who will become our main source of income as we sell and purchase items. We are highly motivated to comply because if he does not let us sell items we can acquire very little money. We can shake small amounts of money out of some trees, and once we have a shovel, we can knock money out of a rock (a different one each day), but we cannot purchase a shovel until we have completed the tasks that the shopkeeper has asked of us. In this way we are forced to go through the 'training period'. As we complete the tasks required of us during our employment, we are given more information on the kinds of goals we can pursue, as well as specific instructions.

> *[Player instigated exchange; August; afternoon; outside; third task: furniture delivery]*
> Gaston:
> > Oh! Is it a half day for you, **Pixel**?
> > So anyway, did you want something from me, or what, mon chou?
> > **Delivery!**
> > **I'm killing time.**
> > **Uh, never mind.**
>
> Pixel:
> Choose 'Delivery!':
> *[inventory page opens; furniture item for Gaston is the only item accessible. Choose 'give this'; it gets handed over]*
> Gaston:
> > Huh? Why are you delivering furniture, **Pixel**?
> > … Hah! So you're **Tom Nook**'s little servant? Have fun with that!
> > If you can't pay up, you gotta work for it! Eh, mon chou?
> > …Well, you ARE lookin' a little sad and pathetic, so I'll help ya out.
> > Here… take my <u>shower</u>!
> > Nahhh! No worries. There'll be plenty of time for you to do me favors later.
> > In the mean time, practice your decorating skills with what I just gave you. Just drop it in your room and then <u>tap</u> your head, mon chou!
> > Then you can <u>slide</u> it around! Just push, pull, or rotate it any way you want.
> > Don't forget to <u>tap</u> your head when you're standing next to some furniture.
> > …Knowing you, that tip will probably be out of your brain in a hurry.
> > So run home and try it before it leeks out of your ears, mom chou!

5.7.2.1.3 Within Lesson Sequences

Repeat the simple to complex progression within each task. Each kind of task begins as a simple procedure, and as the game progresses, players become more sophisticated in how they carry out these tasks. The beginning of the game guides us through the basic procedures we will need to be able to explore and make progress: we begin inside the town hall where the character behind the counter will ask us if we need help. We are told about Tom the shopkeeper, and we are given a map of the town with directions to our house.

> *[NPC INSTIGATED EXCHANGE; August; morning; first arrival (but not player's first avatar); in Town Hall; player's character is named Pixel]*
> Pelly:
> > Well, this is **KAFBURG**'s town hall…
> > So, what is it I can do for you today?
> > I moved in!

    I came by cab!
Pixel:
Choose 'I came by cab!':
Pelly:
    …Oh. Goodness me! You must be **Pixel**! Welcome!
    I'm **Pelly**, and I'm the clerk here!
    <u>Tom Nook</u> already told me all about you.
    So you'll be living with **Kaffy** and **Minki**, is that right?
    That sounds like a lively bunch to live with, now doesn't it?!
    Well, we have many residents who live all around <u>**KAFBURG**</u>.
    Please let me show you where your house is. It's here!
    [map is displayed showing house, circled, and with moving hand pointing to it. Our current location is also shown as a flashing silhouette of a person]
    Well then, **Pixel**… I trust you can use this <u>map</u> to find your new home!
    Oh, wait! Do you know… how to pull your <u>map</u> out?
    What?
    Yep
Pixel:
Choose 'What?':
Pelly:
    To check your <u>map</u>, tap the <u>arrow</u> in the upper-right corner of the <u>Touch Screen</u>.
    Then touch the <u>house icon</u> along the top.
    You can also press the <u>X Button on your Nintendo DS.</u>
    All right, then, good-bye, and please come again!

Once we arrive at our house, we are met by Tom, who then hires us and gives us seven tasks, one at a time to complete. This serves as a basic orientation for how to interact with the townsfolk (we must interact with each one in order to proceed), change our clothing, plant flowers and trees and do a few other things. During this time the residents also tell us about the four main tools we will likely want: a fishing pole, a shovel, a watering can, and a butterfly net. With the exception of the shovel these each have one purpose which is also explained to us.

We learn about most aspects of the game a little at a time: one our first errand involves delivering a carpet to one of the residents. When we give them the delivery, we are given a little information on how we can decorate our homes, but it is not until we have actually changed the appearance of our room that any residents offer further information. We are never told explicitly that we will receive more hints after we try something, but we are sometimes asked by the residents to do favors. Since the mechanisms exist to monitor and record all activities within the game, it is also possible to provide additional detail as it appears to become relevant. Although these tips may be delivered at any time by random chance, they are ALWAYS elicited under certain circumstances, so for example the comment that flowers don't need watering during rain may appear randomly at any time, but one of the residents is certain to approach you to tell you this if it is currently raining.

5.7.2.1.4 Summary

Review at both the micro and macro level. A key aspect of the elaboration theory is its emphasis on the value of timely review. The theory also dictates that systematic review is important, but games are rarely that structured. Nonetheless, review remains an important and integral feature of ACWW in a number of ways. Certain states always trigger certain responses. The game is after all, a computer program and there can only be a finite (and often small) number of ways the game can proceed from any particular point. ACWW uses various mechanisms to trigger review, most of which are delivered through 'conversations' with the other residents, all of whom are NPCs. For example, if you approach one of the residents for a conversation and you have recently been stung by bees, you are likely to experience one of the following exchanges[2]:

> *[player instigated conversation; August; bitten by bees; evening; outside; town fireworks are on]*
> Aurora:
>   Aaieee!!!
>   Whoa! Don't scare me like that, **beefcake**!
>   Sheesh! Look at you. It's like you got stung by every bee in town…
>   Listen, next time, just run straight to the nearest house. Don't u know!
>   Bees can't use doors! In the battle of man versus bee, doors are your savior!
> ----------------------------
> *[player instigated conversation; August; evening; in Purrl's house. I've been bitten by bees]*
> Purrl:
>   And today's doofus of **KAFFBURG** award goes to….
>   Your face!
>   For the esteemed winner, ***Kaffy***, I would suggest some lovely *medicine*!

These two exchanges give us tips on how to deal with bees – they may be repeated whenever the player begins a conversation while showing evidence of a bee sting.

5.7.2.1.5 Synthesis

Make the content structure explicit with visual and other objects. Synthesizers are used to integrate content in a meaningful way and to help learners assimilate prior knowledge. They can be used to organize elements horizontally (relationships among ideas in a single lesson) and vertically ("relationships between ideas in a group of lessons, and the general and inclusive ideas that contain them" (Reigeluth & Stein, 1983, p. 360)). One of the mechanisms used in ACWW that acts as a synthesizer is that of repetitive patterns. While it may have been included for effi-

---

[2] There are in fact more conversation sets that have to do with bees, but these two are representative. The tone of the exchange will always be tempered by the 'personality' of the NPC speaking.

ciency of data space and programming in design, the practice would not continue if it did not also help players. Dialogue sets have specific styles, and players quickly learn what kinds of responses to anticipate. For example, residents will ask the player to answer a question from time to time. The appropriate answer is sometimes rewarded by the NPC giving the player a gift and which answer is appropriate depends on the personality of the NPC asking the question. So players learn to associate certain types of responses with certain types of residents.

[NPC INSTIGATED EXCHANGE; August; early morning; outside]
Bella:
    Oh, hey there! **Beefcake**, I'm thinking of getting a new pet fish.
    So what's a good name for my new fish, eeks?
    Ruby
    Sushi
    Jaws
    Mr. Fish!
Kaffy:
Choose 'Ruby':
Bella:
    Yeah! That's cool!
    I mean, REALLY cool! Thanks, eeks!
    OK, **beefcake**, since you helped me name my new fish, I'll give you this!
    Here!! It's not much, but I really want you to have my <u>modern bed</u>! Love it like it was your own....because it TOTALLY is now, eeks!
[item is exchanged during last statement]

There are also some fairly sophisticated vertical synthesizers that players can recognize and work with, but the choice to pursue the implications of these relationships is always with the player. If this were a game designed deliberately for education, these may be teacher-lead before or even during gameplay. In ACWW, there are many kinds of collections that can be taken up in this game and each is rewarded in different ways, some of which involve relationships not immediately obvious from the game. One of the most sophisticated ones is that of furniture collecting and the interior design of your house. For example, the manner in which players decorate their houses can be left entirely to personal taste, but this aspect of the game, as almost any other has an underlying principle which can be used to improve one's score in that area. Players have their houses rated from time to time by the "Happy Room Academy" - typically after a new item has been added to the home. The rating of the home makes use of several concepts, including furniture 'collections' (all of the same style), special items, and the house's Feng Shui. In this way the arrangement of items within the house can become more than random or player whim - there is a 'system' that can be discovered and followed.

5.7.2.1.6 Analogies

Use analogies that connect with prior learning. If we look at it from the proper perspective, the entire game can be seen as an analogue of activities from real life: earning a living; building relationships; achievement through contests, etc. One of the main functions of analogy is to relate content to the learners' prior knowledge so learners can assimilate newly presented ideas. Another thing that makes analogy a useful device is the ability to create connections that not only reach backwards to what we already know, but also to provide a path for forging forward connections. When we encounter a new situation we will have a frame of reference already built. ACWW does this too - the concept of a mortgage that must be paid off may be familiar to adult home owners, but probably not to younger people playing this game. Through this mechanism, complex and sophisticated ideas can be introduced and players can be guided towards understanding them.

While there are many analogies to real life in ACWW, it must be remembered that this game was intended as an entertainment and not as a serious trainer in life skills. As such, the 'life' analogies are unreliable. Still, part of what makes the game compelling for many players is its mimicry of life. The mechanisms are effective.

On a more detailed level, analogies are especially useful when introducing difficult and unfamiliar ideas. ACWW is not intended to be difficult and as a result there aren't many places where analogies are required for learning within the game.

> *[player instigated conversation; August; bitten by bees; evening; outside; town fireworks are on]*
> *Aurora:*
>   Aaieee!!!
>   Whoa! Don't scare me like that, ***beefcake***!
>   Sheesh! Look at you. It's like you got stung by every bee in town…
>   Listen, next time, just run straight to the nearest house. Eeks!
>   Bees can't use doors! In the battle of man versus bee, doors are your savior!

5.7.2.1.7 Cognitive Strategies

Use a variety of devices to help trigger appropriate processing strategies. Most games employ many cognitive strategies intended to help players discover what they need to know and to remember what they have learned and often games use similar approaches, such as different colors of text to mark specific things. Some of these strategies are so common that they could arguably be considered aspects of basic games literacy. The use of color in text displays is one strategy that is commonly used to indicate classes of words, items, clues and so on. In ACWW examples that follow different types of text styles are used instead of colors. The names of items that can be collected are displayed in *italics*, references to the operation of the game are underlined, names of other residents are **bold underlined**, the name of the town is SMALL CAPS, and the player's name is in ***bold italics***, even if a nickname is being used instead of the chosen name. This way, even if a word is being used for the first time, the player will be able to classify it and thereby know what can be done with it.

> *Tom Nook:*
> If you want to use your <u>*fishing rod*</u>, grab it from your <u>pockets</u>, and you're ready to go, hm?
> ....
> *Pudge:*
> YAAAAAAWN! Good morning…
> Everyone in **KAFBURG** gets up so early.
> I'm **Pudge**… I'm better at wrestling and eating than anybody.

Another common strategy is to restrict the player's options at various times. When trying to sell items at Tom Nook's store, for example, only those items Tom is able to buy are accessible in the inventory even though the player may have other items in the same place. The others are still visible, but appear faded (i.e. in the background) and they cannot be grabbed or moved. In this specific situation inaccessible items include such things as money, which Tom does not buy. In other situations the subset of accessible items will be different, and occasionally the items may be 'grab-able' but cannot be used. It is possible to attach items to letters we may send to other residents as gifts, but some kinds of items (like fish and bugs) can not be attached and simply will not 'stick' to the letter if dragged over to it.

Suppose we decide our primary goal in the game is to pay off the mortgage. The first mortgage amount is 19,800 Bells, and at the start of the game the player has no tools so there are few ways to earn money. If we simply wander around, various residents will tell us that we can shake fruit out of trees and pick up shells at the beach. Tom the shopkeeper will buy whatever we have collected and as soon as we have picked up enough items to buy some of the tools, we can begin to fish, catch bugs, and so on and we are well on our way.

5.7.2.1.8 Learner Control

Encourage learners to exercise control over both content and instructional strategy. Because ACWW is an open-ended role-playing game, one would expect to have much latitude when it comes to exercising control. Player/learner control is possible in many places in the game and at many levels of abstraction, starting from the look and feel of the game and the main character, to how the NPCs sound, but the greatest latitude in this game is afforded in the game play itself. Players may choose to focus on as many or as few aspects of the gameplay as they like. They may decide to focus on collections, making money, interacting with the other residents, or even fashion design.

Text is displayed in small chunks, but users are given the option of scrolling through text faster if they want. Most dialogue is given to the player no more than a dozen words at a time, at the end of which the user must use a control to go to the next bit. Another way that many games provide control is with the appearance of the player's avatar. ACWW players may choose their character's gender as well as various other things before the game begins, but in this game the process is presented as part of the game's introductory sequence rather than a pre-game activity. At the beginning of the game while the character is riding to town in a cab, the taxi driver asks the character various questions and makes various comments. Players have the opportunity to respond at several points, and those responses determine the character's gender, eye color, hair style, and several other things. "The next question asks how you like the name, and is the first step toward determining your character's gender. If you select "That's not it" then you can re-enter a new name in case you made a mistake. If you tell him you think the name is cute, he'll think you are a girl. If you tell him it's burly then he'll think you are male. After this question you will have the opportunity to select "I'm not a boy/girl!" if you chose the wrong one, don't worry." (Eagleson, 2006)

5.7.3 Merrill's First Principles of Instruction

M. David Merrill's career in instructional technology has spanned 40 years, and includes numerous significant contributions to the field. He is probably best known for his component display theory (Merrill, 1999). In the 1990's he was one of the foremost proponents of "second generation instructional design (ID2)" (Merrill, Li, & Jones, 1991b) which acknowledged a more open-ended and less prescriptive approach to ID, and included Instructional Transaction Theory (Merrill et al, 1991a) and ID based on knowledge objects.

The first principles of instruction are the result of a systematic review of instructional design theories, models and research. Each of the principles included satisfies the following properties:

- promotes more effective, efficient or engaging learning,
- is supported by research,
- is general enough to apply to any delivery system or instructional methodology, and
- is design oriented with direct relevance to promoting learning activities.

Merrill defines a 'principle' as a basic method, and describes it as a "relationship that is always true under appropriate conditions regardless of program or practice (variable methods)." (Merrill, 2002, p. 43) There are five principles that constitute a set of fundamental elements common to all effective instructional design. Merrill hypothesizes that 1) "Learning from a given instructional program will be facilitated in direct proportion to the implementation of first principles of instruction", and 2) the "learning from a given instructional program will be facilitated in direct proportion to the degree that first principles of instruction are explicitly implemented rather than haphazardly implemented." (Merrill, 2002, p. 43) If true, then illustrating connections between game elements and Merrill's first principles would suggest that games facilitate learning in substantial ways (See Figure 5.3).

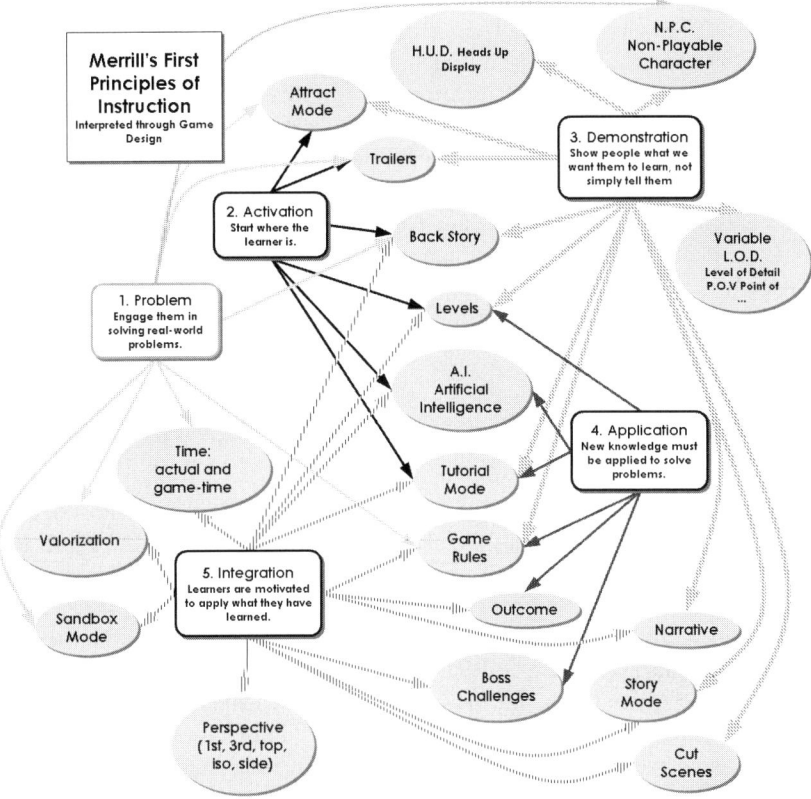

Fig. 5.3 Connections between Merrill's First Principles and Common Game Elements

5.7.3.1 Merrill's First Principles of Instruction Starring Mario

5.7.3.1.1 Problem

Engage them in solving real-world problems. Obviously, claims that the actions within The New Super Mario Bros. emulate any sort of real-world problem other than an obstacle race would be a stretch. However, if we take it to a higher level of abstraction, the challenges of practice and rewards associated with incremental progress could be described as real-world problems. In this case however, I would argue that the real-world nature of the problem is less important than that it be engaging, and for millions of players, Mario is certainly that. Parables, fairy tales, and fantasies can all be engaging and can all relate, even if only indirectly to real-world problems. Again, in this example it must be remembered that the object of this study is not so much to discover how this game is educational as it is to discover how the mechanisms used in this game embody the educational principles outlined.

5.7.3.1.2 Activation

Start where the learner is. In most successful games trailers, back stories, and tutorials all contribute to helping the player become familiar with the game and its basic rules and objectives. Being a commercial enterprise the designers would naturally want to attract as large an audience as possible, so many 'first' games assume little more than basic knowledge of the game equipment, whether it be a console, handheld device, or a PC. Some 'numbered' games (Final Fantasy XII, Call of Duty 3, and other sequels) assume prior knowledge through having played previous editions of the game, but those that rely on this too heavily also restrict their audience and that is rarely desirable in a commercial context.

Each *Mario* game has similarities with each other, but players need not know anything about any one of them to play and enjoy any other. Successful games almost always have simple challenges at the beginning that shift more or less gradually to ones that can be surprisingly difficult. Mario allows players to replay any level they have already completed, providing a considerable degree of flexibility and the option of returning to 'familiar' territory to regain confidence before attempting a particularly tricky course.

5.7.3.1.3 Demonstration

Show people what we want them to learn, not simply tell them. In an effort to ensure that learners come away from our lessons with all those things we feel are fundamental or essential, we sometimes provide too much demonstration. This is especially true when the learners are gamers as they are accustomed to trying things out after only minimal tutoring. After all part of the fun in a game comes from discovering unexpected places, treasures, and how to do things while poking around and learning how to play the game. Learning is not usually efficient, and the commercial game designer's desire to keep players engaged with their games as long as possible supports that notion well.

There are a great many devices used to indicate what players need to learn in order to win a digital game. It is assumed that we understand the basic premise, and this is facilitated through the use of an introductory cinematic clip. In this game, we see Mario and Princess Peach out for a walk together, a disturbance at the distant castle distracts Mario and while he is away investigating Bowser sneaks up and snatches the Princess away. In the opening cut scene of the first level we catch a glimpse of Bowser carrying Princess Peach into the tower which is also one of the courses we must complete, but from where we are there is no path by which we can reach it. Thus, we conclude that we must somehow build or generate a path to that tower so we may have a chance to go in and retrieve her. Basic games literacy includes the knowledge that moving, flashing, or otherwise highlighted items in a game are almost always significant, so these should be hit, shot, picked up or otherwise acted upon. It also includes the sure knowledge that there IS a way to win. In *Mario*, the way to generate new pathways is to enter each course, and get to the end while collecting as many points as possible. Along the way we are shown star coins we can grab, and hints at hidden pathways. This is a game after all, so it is assumed that the hidden pathways will reveal something of value. A typical game is expected to offer twenty or more hours of gameplay, so any game that demonstrates too much of what we need to learn and does not allow us enough opportunity for discovery is unlikely to become successful.

5.7.3.1.4 Application

New knowledge must be applied to solve problems. The whole point of a game is to meet some challenge or solve some problem, preferably many of them. Given that Mario lacks a rich back story, the puzzles and challenges are entirely the point of this game. Each level offers several new challenges as well as increasing levels of difficulty on previously mastered skills. The first level of any 'world' will introduce the basic maneuvers used throughout that set of courses. After discovering that some blocks release points and others release power-ups one of the first things we will do when we come across a block is to hit it to see what happens. In doing so we also discover (usually too late at first) that some blocks break away when hit, which may deprive us of a needed ledge or jumping off point.

5.7.3.1.5 Integration

Learners are motivated to apply what they have learned. Each course also presents the player with new situations and challenges that call for application of what has already been learned, but in new ways. The set of 'skills' required to survive to the end of the first course is quite small, but each time we proceed on to the next course we bring those skills with us. Having learned that we have some control over the speed at which Mario runs, we encounter a log we must cross that teeter-totters. The motion becomes exaggerated as soon as we jump on and so we must move at the appropriate speed to get across without tipping it too far and sliding off. One of the hallmarks of successful games is that they almost always provide the player with many opportunities to practice and apply any skill learned within the game. In fact any time a new skill is acquired in a game it is assumed to have some additional purpose later on. This assumption provides considerable motivation to players to spend both the time and the effort to acquire skills, and a game that requires players to learn skills and gain knowledge that are then never used again tends not to remain popular for long.

5.8 New Frontiers

The will to learn is an intrinsic motive, one that finds both its source and its reward in its own exercise. The will to learn becomes a "problem" only under specialized circumstances like those of a school, where a curriculum is set, students are confined, and a path fixed. The problems exist not so much in learning itself, but in the fact that what the school imposes often fails to enlist the natural energies that sustain spontaneous learning.

(Bruner, 1966, p.127)

This next section looks at a few less structured learning and instructional theories using the same games as were used in the previous section. If the same game can be seen to embody sound pedagogy even if studied through multiple lenses, then that will lend additional weight to the main thesis of this chapter.

The author wishes to apologize in advance if I have not included your favorite theory or model here. I have largely avoided the social learning theories as for this effort I am looking JUST at the game and how its design stacks up against accepted pedagogy rather than the dynamics of communities of players. Even so, it must be acknowledged that any game that includes NPC's provides a community of a sort, and although all possible behaviors of these NPCs were deliberately designed the effect is still one of social interaction.

Some theories and frameworks do not lend themselves to analysis in the way this chapter proceeds. Others, like legitimate peripheral participation, situated learning, and apprenticeship are important concepts and highly relevant to learning in game environments, but they do not really provide a clear framework that we can use to develop evidence if our goal is to build an argument that the medium of the game is a legitimate educational technology. Legitimate peripheral experience is the central defining characteristic of how newcomers become part of a community of practice. Knowledge is situated and therefore effective learning should take place within the context in which the knowledge will be applied (Lave & Wenger, 1991). Aside from using these comparisons to argue that commercial games already employ sound instructional design principles, the theories chosen here can also form the basis for the kinds of design principles that will be of use as we move forward and begin to try and teach others about educational game design.

5.8.1 Activity theory

Activity theory is not new, having been developed in the early part of the last century by Lev Vygotsky (Vygotsky & Cole, 1977), A.N. Leont'ev (Leont'ev, 1978), and A.R. Luria (Luriëiia, 1976) in Russia. The main focus of this theory revolves around the interrelationship of the subject (the learner), the object (the goal which leads to the outcome), and the tools (both physical and conceptual) used to mediate between them. It suggests that the relationship between objects in the environment and people are mediated by culture and its rules, the community, and by labor and its roles and development.

Others have already applied this theory to games (Dobson, Ha, Mulligan, & Ciavarro, 2005; Hadziomerovic & Biddle, 2006; Oliver & Pelletier, 2004; Squire, 2002) and have studied it in the context of player learning. The current effort examines games and activity theory from the perspective of the game design (See Figure 5.4) and in that respect also expands on work in human computer interaction (Kuutti, 1996). Very loosely described, in this view of activity theory, the subject is thought to form a relationship with the tool, but that the tool only becomes a tool through the user's activity. While the current examination cannot fully detach the user's relationship with the game, the focus here is more on the design of the artifact (the game) and how that design embodies the concept of this theory. Also, Kuutti (1996) has defined three levels of analysis that are also useful to examination of games through activity theory (Pelletier & Oliver, 2006), namely: strategic level activities (such as paying off a mortgage or completing a collection), tactical actions (like catching a fish or gathering fruit off a tree), and operational actions that follow a particular pattern that can become automatic (like giving Tom Nook items you wish to sell from your inventory, or like fishing) so long as nothing goes wrong causing a contradiction (like being bitten by a mosquito while fishing).

Activity theory is descriptive rather than predictive, and as such offers a useful perspective through which to view the design of games.

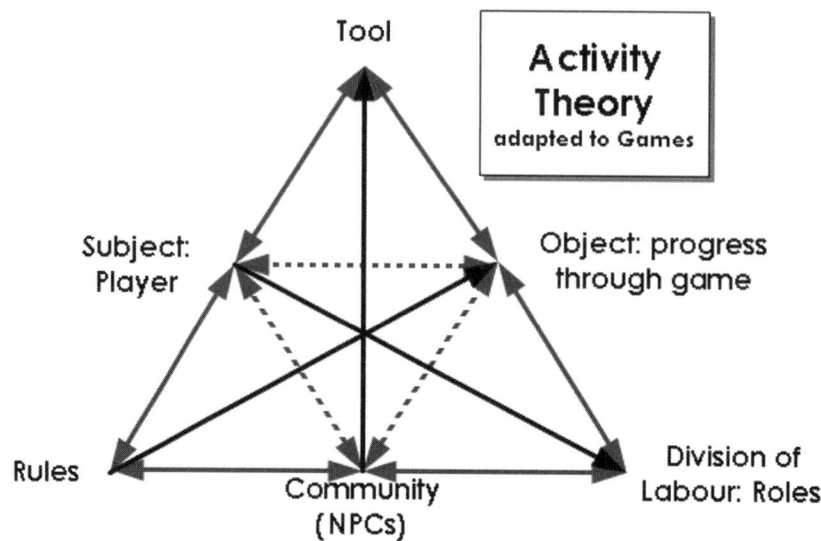

Fig. 5.4 Activity Theory Adapted to Games

5.8.1.1 The Activity Theory of Animal Crossing

5.8.1.1.1 Subject

The player is the subject. The main subject of any game is the player of course, and in ACWW players may take on several characters, but not simultaneously. As many as four player/residents can 'live' in one ACWW game but whenever one is awake the others will always be sleeping. These other characters are still distinct in terms of assets like money and possessions, and their relationships with the other residents. This game is played from a third-person perspective so the player, while likely identifying with their character, sees that character act within the game. This perspective adds a degree of distance for the player as the play experience is more akin to playing with a doll than pretending to be someone else.

5.8.1.1.2 Object

What is the objective? In ACWW the choice of object affects how the game will best be played, although this game does encourage some activities more than others. The main ones include: relationship building, collections, money making, the "stalk" market (the currency of trade is turnips), and gardening (which includes fruit and money trees as well as flower breeding that can result in the production of various new colors). Different approaches are appropriate for different objects, but they are all introduced early in the game, after which the player is free to focus as desired. Players need not commit to any specific goal and may change their focus as often as they wish.

5.8.1.1.3 Tools

What are tools that help the subject achieve the objective? Tools serve as mediating elements in any activity and can be physical, conceptual or symbolic. They include instruments, signs, procedures, machines, methods, laws, and forms of work organization (Jonassen & Rohrer-Murphy, 1999). If we stretch the notion of physical to include in game artifacts with which the player can interact, then all three exist within many games. ACWW has a great number of artifacts and objects but not many classes of objects, and even fewer actual tools that can be used to achieve goals. There are six primary 'physical' tools in ACWW: a shovel, a fishing rod, a watering can, a butterfly net, an axe, and a slingshot. Examples of conceptual tools include: humor (Dormann & Biddle, 2006), relationships that develop between the player and the NPCs with which we interact and over which we have varying degrees of influence, and the use of time, which can be considered a mediating tool as well as having ties to the games rules, and will be discussed further in the next section. At its most fundamental level, the entire game is a symbolic tool but that does not help us further our analysis so we look in the game. Here we find that the symbolic tools include such things as special events, changing seasons, lucky furniture items, silhouettes of fish seen underwater, and so on. All of these artifacts can be used by the players to support progress towards the object and outcome.

5.8.1.1.4 Rules

What the rules by which the subject must abide? Among the defining characteristics of any game are its rules to the extent that a game without rules may not even be termed a game, and in a role-playing game, those rules tend to be fairly complex. ACWW has both explicit and implicit regulations, norms, and conventions that constrain individual action and group interaction. There are positive or negative consequences to almost every action, although in this game the connection between the act and the consequence is often not direct. There are for instance ways to increase one's 'luck' when fishing, which include placing lucky items in our house. Fish appear at random according to a predetermined probability but this can be affected to a certain extent by our own actions. Lucky items placed in our house increase the likelihood of the appearance of rare fish.

Rules of interaction both with the game environment and with the NPCs are enforced in ACWW largely by restricting the user options and also by the way in which the NPCs respond. If we ignore or deflect requests for interaction by residents too often they are likely to stop giving us gifts for example. Residents will move out of our town from time to time, but how we respond to them will often affect when this happens. Since there is no single win state in this game, there is also no single lose state, and the game can continue indefinitely providing essentially endless opportunities to try again even without restarting the game.

In ACWW, game time is intended to match real time and certain events take place on a regular basis (weekly visits by occasional characters, daily replenishment of the shopkeepers stocks, annual special events, etc.). These events are tied to the game's calendar and clock and although players can adjust the clock forwards and backwards, each day that is 'skipped' still exerts influence. For example a small number of new weeds will grow each day (we are supposed to help keep our environment nice by tending the area), if we skip ahead too many days we may find our town overrun with weeds and that the flowers have wilted away. Also the lack of interaction with the residents will prompt many of them to move away.

Rules underpin all progress in digital games.

5.8.1.1.5 Community

What elements and actors form the community for this activity system? All three games being examined are being considered only in the single player mode as it is the design of that game itself rather than the social interaction of various human subjects that is the focus of this discussion. Given that, the individuals or subgroups who share the same general object include only those NPCs who are "on my side" or who are designed to assist the subject rather than hinder her. In ACWW there are a total of twelve regular residents (such as Tom Nook the shopkeeper), 16 occasional visitors and 144 villagers, each of which has one of six distinct personalities (Eagleson, 2006). The town regulars are largely benign and will help out according to their roles, while the occasional visitors are largely beneficial (although a few are scoundrels!) and each of these also has specific roles as well as peculiarities. Of the 144 villagers, a maximum of eight may reside in your town at any given time. Conversations are not free form and in fact very few modern games offer anything but the most rudimentary forms of language recognition. Most conversations involve a pre-determined (or randomly selected) phrase that is displayed, followed by several potential responses from which you may choose.

5.8.1.1.6 Division of Labor

How are the roles delineated in this activity system? The division of labor in a game activity system comes from the ways in which the community is organized. In most games the division of tasks between members of the community is quite well-defined and it is not uncommon for individuals or groups to exist specifically to serve tightly defined roles but tend to be more mobile. In ACWW the regular residents have roles associated with a specific space. Occasional visitors have similarly specialized roles. Lyle the insurance salesman (a weasel) appears once a week solely to sell insurance, and hangs around out house. Tortimer the Mayor for example is only found outside the town hall, and ONLY during special occasions. Blathers the owl can be found only in the museum. In games this mechanism helps to compartmentalize the behaviors and possible actions, thus controlling the game design's complexity while at the same time allowing for player flexibility. From a learning perspective each character and location becomes associated with specific activities and acts as a mnemonic that players can remember to provide a scaffolding effect.

For the player, activities and roles can change as the tools do, but these roles are the ones that the player decides to take up, whether it be gardener, collector, fashion designer, or what have you. In each case, certain game characters and tools will become more significant while others become less so. Regardless of the player's goals though, the roles of the NPCs and other artifacts rarely change in this game.

5.8.2 Constructivist Learning Environments

The notion of constructivism is by now well-known and should no longer require a lengthy explanation. The fundamental view behind constructivist learning environments is that technologies can and should be used to keep students active, constructive, collaborative, intentional, complex, contextual, conversational, and reflective as illustrated in Figure 5.5 (Duffy, Lowyck, & Jonassen, 1993). To many, suggesting that most modern game environments are inherently constructivist learning environments will not come as a surprise. Still, it is one thing to say they are and another to show that they are through explicit connections. A brief explanation of each of the major elements of such an environment is offered in the next few paragraphs, followed by more detailed examination of one game as seen through this lens.

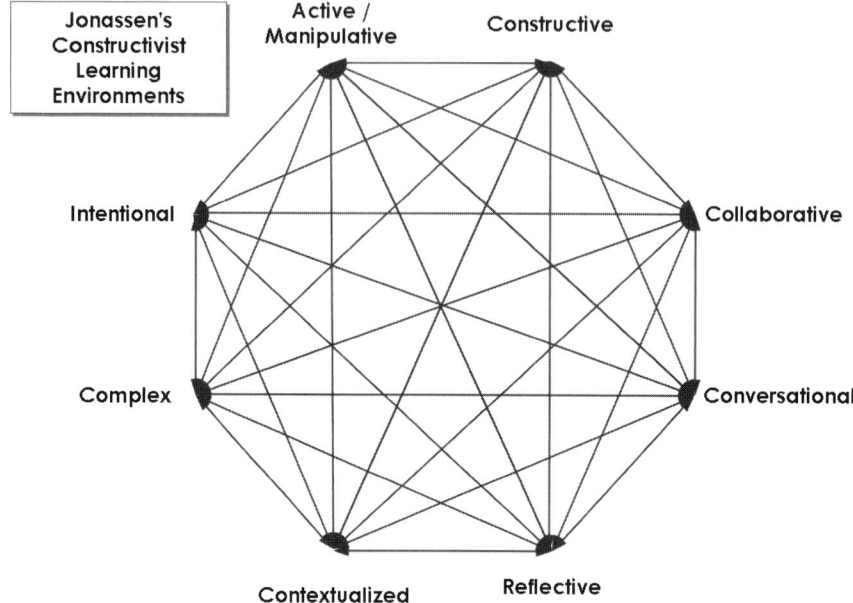

Fig. 5.5 Constructivist Learning Environments

5.8.2.1 The New Super Mario Bros. as a Constructivist Learning Environment

An ideal constructivist learning environment would give the learner a great deal of freedom to interact with it and still present the learner with interesting problems to solve and things to discover. Although Mario is an older style platform game played on a two dimensional plane, it still holds up well when scrutinized as a constructivist environment. In fact the dominant teaching mode, if we can even call it that is through trial and error or discovery learning. While it has already been admitted that Mario hardly qualifies as an educational experience, it is still an excellent example of a game that has remained popular for several decades. It is a game people of all ages willingly choose to play, while *Math Blaster* (Davidson, 1986) a well-known educational game that employs essentially the same style has only enjoyed popularity in schools. Given that contrast, an examination of the mechanisms employed in Mario might provide some hints as to what we should be including in our educational games to make them more engaging.

5.8.2.1.1 Active

"Learners are engaged by the learning process in mindful processing of information where they are responsible for the result" (Jonassen). This notion lies at the very heart of most digital games, and Mario is no exception. There are very few places in each level or course where the player can let Mario stand about and do nothing. This is often cited as one of the great attractions of this game: the game is fast-paced and players are always busy. Various villains are also present in each course and since they move around as well, Mario must be vigilant and ready to act. Mario has almost no dialogue beyond the occasional, "It's Mario!", "Bye-bye!", "Here we go!" and various onomatopoeia, and the music is cartoon-like and up-beat, as is the imagery. Each level is associated with a specific 'world', and when the game begins we have access to only a single level on the first world. All other levels must be earned through our satisfactory performance. Success in this game is all up to us, and there is no way to get through the levels except by practice and more practice.

5.8.2.1.2 Constructive

"Learners integrate new ideas with prior knowledge in order to make sense or make meaning or reconcile a discrepancy, curiosity, or puzzlement" (Jonassen). The format of the game is essentially the same as a traditional obstacle course - there are a total of 80 different ones in eight different 'worlds', each of which has a different type of landscape - one is made up of deserts and another is entirely underwater. The courses in each world have a similar look and feel to each other with some differences as well and each one builds on some skills learned in a previous world while adding one or more new challenges. The first level introduces us to all the basic skills and 'power-ups': here we can learn to jump, run, and dash. We learn how to break blocks, catch star coins and we meet several of the 'classic' Mario bad guys: Koopas and Goombas. When viewed from a different perspective the game is constructive literally as well as conceptually in that we begin with access to a single level in a single world and we are given the impression that other levels are there but the pathways needed to reach them do not yet exist. Once we have successfully made it through to the end of the first course the next section of pathway is constructed and we gain access to the second course. Access proceeds in this way with more and more of Mario's world becoming accessible as more and more pathways are constructed. Once we have made it through the final course in the current world we gain access to the next world.

5.8.2.1.3 Collaborative

"Learners naturally work in learning and knowledge building communities, exploiting each others skills while providing social support and modeling and observing the contributions of each member. Humans naturally seek out others to help them to solve problems and perform tasks" (Jonassen). The current analysis only looks at the single player version of this game and since Mario acts alone and all of the NPC's are bad guys in this game there is little opportunity for collaboration within the game itself. Outside of the game, however there exists a thriving community of Mario fans eager to collaborate on anything from fan art and fan fiction to sharing tips, techniques and even videos of gameplay. Collaboration and the communities that sustain them are very strong with almost all popular and successful games.

A casual search on the web using the phrase "New Super Mario Bros." turned up 1,600,000 hits! Almost all of the first page of links (50) were sites offering reviews, previews, cheats, walkthroughs and hints, so there is clearly no shortage of players keen to share what they have learned and add to their knowledge with the help of others. One walkthrough guide is produced in full color and is 87 pages long (Sallee, 2006)!

5.8.2.1.4 Intentional

"All human behavior is goal directed (Schank & Cleary, 1995) That is, everything that we do is intended to fulfill some goal. When learners are actively and willfully trying to achieve a cognitive goal (Scardamalia & Bereiter, 1994), they think and learn more" (Jonassen). The motivational power of fun and humor should not be undervalued (Dormann & Biddle, 2006), and Mario lacks neither. The sights and sounds are amusing and Mario's reactions are pleasing to the extent that repeating challenging portions of a course over and over until we get it right sustains us. The response of the game when we finally reach the end of a course is also appealing, and we are often additionally rewarded with the release of additional courses. A game that has very little point beyond entertainment MUST succeed here, or players will stop playing. Even worse for the game makers, they will not encourage their friends to purchase the game. Though the goals may be trivial, players will always be able to identify the goal they are trying to fulfill at any given point in this game.

5.8.2.1.5 Complex

We need to engage students in solving complex and ill-structured problems as well as simple problems (Jonassen, 2004). *Mario* is largely a game of skill and although there are few complex intellectual conundrums, there is no shortage of complex puzzles. Here is a brief description of how to proceed through a portion of the level five tower course:

> "Avoid the two Spiked balls that are rolling around here, getting the ? Block Power-Up, with the following ? Block set (the one on the right) holding a 1-Up Mushroom. Immediately after this portion get Mario towards the right side of the ledge as a Giant Spiked Ball will destroy the bricks on the left, signaling the point for Mario to start hopping up the ledges ahead, keeping himself above the nasty implement. Keep hopping upwards, fading to the right to get the third Star Coin on that side, followed by getting onto the ledges to leap upwards as soon as ledges appear because that Giant Spiked Ball will get back to the right side soon enough. Punch the bricks on the right to gain one last Power-Up, followed by passing through the large red doors to encounter the boss fight!" (Sulpher, 2006)

Mario presents players with challenge after challenge, some simple enough to meet on the first try and others complex enough to drive all but the most dedicated players to the game playing community for help and hints.

5.8.2.1.6 Contextual

Learning should be situated in some meaningful real-world or case- or problem-based task. Let's face it, Mario bears little resemblance to any real-world activity, but it certainly presents challenges. In fact, it is the contrived nature of the entire game that makes it an interesting design to study from an educational perspective. Although we are often told how important it is that the game's premise and story-line be fully integrated into the gameplay in order for the game to be 'good', Mario's story is very weak and most of the activities we must master really have nothing to do with rescuing Princess Peach. Mario is a series of obstacle races so there really is no meaningful context to speak of. Yet, it still works and Mario remains one of the most popular and recognizable characters of all gamedom. Some of the reasons for this enduring popularity are probably similar to those that fuel the popularity of other cartoon characters like Mickey Mouse and Bugs Bunny, but part of it has to do with the integrated style of the Mario games: all have a similar look and feel; nothings seems out of place. This is where context plays an important role in an otherwise meaningless collection of silly activities.

5.8.2.1.7 Conversational

"Learning is inherently a social, dialogical process (Duffy & Cunningham, 1996). That is, given a problem or task, people naturally seek out opinions and ideas from others. Technologies can support this conversational process by connecting learners across town or across the world" (Jonassen). Here again we turn to the wider game community because Mario is a single player game. Mention has already

been made of the role played by the internet game communities while discussing collaboration, and it will be mentioned again in the next section, but one aspect that has not yet been included is that of other people likely to be in the same room as the player while they are playing. Mario can be a popular spectator sport. Mario also supports a two player mode which further encourages social interaction during play.

5.8.2.1.8 Reflective

"Learners should be required by technology-based learning to (articulate) what they are doing, the decisions they make, the strategies they use, and the answers that they found" (Jonassen). This is one place where many games are somewhat lacking in design (Prensky, 2001). However, virtually all popular games spawn online fan communities whose main purpose is to allow players to share tips, knowledge, experiences, and artifacts that they have produced, such as fan art and fan fiction. If anything, in these venues popular games suffer from an overabundance of reflection. While they may not structured in ways educators find appealing, they certainly serve the same purpose, namely to reflect on the experiences of playing the game and discuss what they have learned.

5.8.3 Problem-based learning

Problem-based learning (PBL) is intended to build on the efficacy of experiential learning and promote learning through an investigation of a problem which learners must solve in groups or individually, in role-playing or scenario based contexts. Learning is student centered and relies upon self-directed learning (Savin-Baden, 2000). When designed as a formal exercise, PBL includes an introductory exploration of issues, followed by a development of the problem to be solved. It is usually assumed there will be a collaborative group of participants involved in the process who then hypothesize about possible solutions, gather information needed to resolve the problem and then present their solution when they are done (See Figure 5.6).

Let us now follow along as Phoenix Wright and his friends solve one of the problems with which they have been saddled. Each of the five cases in this game can be viewed as a main problem, to which the PBL process can be applied. With each main problem there are also lesser problems and focusing this analysis on one of these provides a fairly complete picture of the process. It also helps to illustrate that the PBL process can be nested within itself in a procedural fashion: there can be one main problem which contains other problems, each one of which can be addressed using the same process.

The design of this game itself matches very closely with the formal PBL approach.

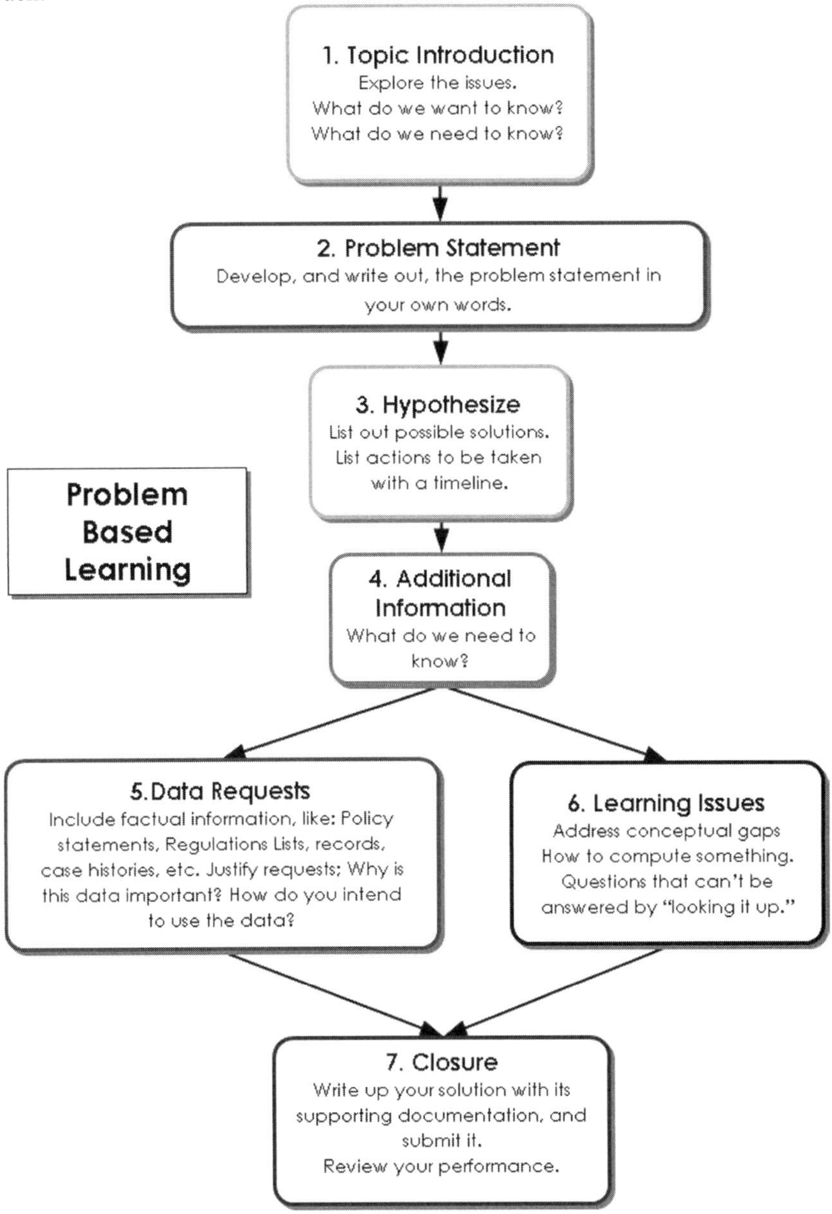

Problem Based Learning

1. **Topic Introduction**
Explore the issues.
What do we want to know?
What do we need to know?

2. **Problem Statement**
Develop, and write out, the problem statement in your own words.

3. **Hypothesize**
List out possible solutions.
List actions to be taken with a timeline.

4. **Additional Information**
What do we need to know?

5. **Data Requests**
Include factual information, like: Policy statements, Regulations Lists, records, case histories, etc. Justify requests; Why is this data important? How do you intend to use the data?

6. **Learning Issues**
Address conceptual gaps
How to compute something.
Questions that can't be answered by "looking it up."

7. **Closure**
Write up your solution with its supporting documentation, and submit it.
Review your performance.

Fig. 5.6 Problem Based Learning

5.8.3.1 Problem-Based Learning with Phoenix Wright

5.8.3.1.1 Topic Introduction

Each case begins with a formal topic introduction. The fifth and final case is the most complex and twisted of all, which should not be especially surprising as we expect a progression of difficulty in almost any game we play. The synopsis of the case is as follows: the district chief prosecutor is accused of murdering a police detective in the underground parking lot of the prosecutor's office building. There is a witness (Angel Star, a former detective) and the chief prosecutor (Lana Skye) has confessed to the crime, but Lana's 16 year old sister is certain she didn't do it and asks Phoenix Wright to defend her big sister. We already know that we must prove her innocent and find the real murderer because that is how this game is played. The cases, although they start off being relatively straight-forward quickly become quite convoluted.

The 'sub-problem' we will use as an example is one that occurs at the end of the first day of the trial. This case is divided into six chapters which alternate between evidence gathering and in-court trial episodes. The game is organized such that when a new problem arises, the judge will call a recess and we are given an opportunity to gather more evidence and uncover more information. Once we have the information and evidence we need we go back into the courtroom. This change from one chapter to the next is controlled by the game, and we cannot choose to go back to a previous chapter once we have passed it.

5.8.3.1.2 Problem Statement

The specific problem that we will examine occurs when during a witness testimony (that of Damon Gant, The Chief of Police) we discover that there had apparently been a second murder inside the Police Department on the same day and at the same time as the one incident of the current trial. The body was not found, and there are indications that a case that was resolved two years ago might somehow be linked to this one. Evidence from the two-year old case was found at the scene of the current crime but has not been shown to be linked to the current victim. The Chief of Police claims that there is no official link between the two murders that occurred on the same day. Through the evidence presented, we manage to prove the connection between the two current murder cases, but there is still no second body. Further testimony and cross-examination shows that the murder victim in the second case appears to be the same person as the victim in our current case. Our problem is now to clear up the mystery of how the victim could have been killed in two places at the same time.

5.8.3.1.3 Hypothesize

The next step in the PBL process involves hypothesizing, and in the game this is accomplished by dialogue exchanges between various characters at the start of the evidence gathering chapter. As players, we have very little to do in this section except to scroll through and read the dialogue. Through the dialogue we are given a brief review of what we know, told what we think has happened and also told what we should do next in order to resolve the problem. In the story, we have two murders that occurred at the same time but there is only one body – this is not very likely. The accused (Lana Skye) will not tell us anything – we suspect she is trying to protect someone or something. Given the nature of this game, we (as players) can hypothesize that most of the information given to us is significant in some way, so it is reasonable to assume that the references to the two-year old case are significant, and in case we didn't get this on our own, we are given hints to that effect. The game essentially points us at the next phase of the PBL process, namely identifying what additional information we will need in order to proceed.

5.8.3.1.4 Additional Information

One of the game mechanisms added in this case is the ability to examine evidence by rotating the image of the object along two different axes as well as to zoom in and examine certain portions of the object more closely. In the previous cases the only things we could do with evidence were to look at a picture of the evidence and read a few details. If the evidence happened to be a piece of paper (note, letter, autopsy report, etc.) we had the ability to read whatever excerpts the game designers decided to put in for us).

The evidence gathering portion of the game is facilitated through a small number of locations we can visit. Most cases have fewer than 10 different locations. At each place, we have at least two options: 'examine' or 'move'. The 'examine' mechanism was described in a previous section (Gagné's), so it will not be described again here, except to add that it is through this mechanism that we can find and collect evidence[3]. The move option takes us from one location to another, but we can only move along certain pre-determined paths, so for example, if we wish to go from the Police Department to the High Prosecutor's Office, we must first go to the Underground Parking Lot, because the prosecutor's office is only accessible from there.

If we arrive at one of the locations and there is another character there, the game will provide two additional options: 'talk' and 'present'. These two options are the mechanisms by which we fulfill our data requirements and our learning issues. Just as in other PBL exercises, it is not always easy to decide which one is a

---

[3] The other way to collect evidence is when it is explicitly given to us by one of the other game characters.

data requirement and which should be labeled a learning issue. In the game there is typically some initial dialogue followed by an opportunity to instigate further 'talks'. All 'talks' are necessary in this game, and sometimes additional ones will become available after others are done – this is one way of keeping the player from seeing hints that they would not yet understand. The 'present' option allows us to show evidence to the character to see if it can provoke the character into giving us more information. We always get the same response if the witness finds the evidence unmoving, but there is no penalty for trying.

5.8.3.1.5 Data Requirements (facts)

Another way of looking at the distinction between data requirements and learning requirements is to say the one deals with 'what' while the other addresses 'how'. In the case of "Rise from the Ashes", the data requirements will include the facts and clues of the case that we will need to gather in order to resolve the problem. In the course of gathering the data we need we will also encounter learning issues, and vice versa, so it will rarely be possible to group items neatly into one or the other category. While they can be listed separately after the fact, since this example follows the process as it is unfolding, the remainder of the discussion follows under the next section.

5.8.3.1.6 Learning Issues (concepts)

In addition to concepts, learning issues include the 'how' that goes with the 'what'. So for example, in order to determine the facts of the two apparent murders, we will want to examine both murder scenes – if we can find conflicting clues, we will be able to pursue those. At the start of the evidence gathering chapter, we find ourselves talking to Ema Skye, the defendant's younger sister. She mentions something about blood stains, and presents us with a 'luminol kit' – a spray that creates a stain when mixed with blood – even trace amounts. One of the learning issues we encounter is how to use this new device, and we are given a short tutorial on the spot. Then, when we use this kit we discover that there are multiple blood stains at the scene of the 'mystery' crime and only one at the scene of the current one. After various other talks and examinations, we realize that the evidence suggests that there was only one murder and that somehow the body was moved from one location to the other, which explains the missing body.

5.8.3.1.7 Closure

Rather than spoil the mystery, let it just be said that we eventually find all the evidence we need. This includes the addition of yet another tool – a fingerprinting kit.

This new tool takes advantage of the touch sensitive screen and built-in microphone of the DS, and allows us to choose a suspect print on some surface, dust it by touching the screen, and then literally blow away the excess dust revealing the print. As soon as we are given this fingerprinting kit, our set of character profiles is altered to include their fingerprints so we can compare what we have against our records. The final step in this process is to suggest a match whereupon the game mimics a fingerprint matching program by focusing of several key points on the fingerprint for comparison. This tool illustrates an innovative use of the interface as well as adding a flavor of authenticity.

Ultimately, all the evidence we need is gathered and we can press our witnesses by asking further questions during the cross-examination of their testimony and by presenting that evidence at the appropriate moments. Since this examination was of a sub-problem, the closure of this problem allows us to continue on to the next chapter, and ultimately prove our client's innocence. The format used in *Phoenix Wright* fits very cleanly with the PBL format, and even though this game is essentially a branching story where we are led towards the end in a fairly lock-step fashion, the interactive tools we are given and the opportunity to 'solve the case' ourselves makes for a compelling and enjoyable experience. The genre of the mystery is one that could be used in a great many learning situations, and this game provides a template for how to translate that into a game format, while still retaining the control that might be appropriate in certain learning situations where a 'right' answer must be the ultimate conclusion.

5.9 Digital Games Are Special (Educational Technologies)

I have learned throughout my life as a composer chiefly through my mistakes and pursuits of false assumptions, not by my exposure to founts of wisdom and knowledge.

Igor Stravinsky

None of the games we looked at are educational but we can still see that the templates used by each game maps very nicely onto several well known and respected learning and teaching theories. The medium of the video game supports many traditional approaches to learning as well as many modern ones and some would say it does so better than many other formats. Whether we are looking to develop a full-blown game or just some aspect of game technology, we can learn from the masters.

Phoenix Wright may not teach us anything realistic about being a lawyer or solving crimes, but it is a compelling format for dealing with mysteries, and we could just as easily use this approach to work through an environmental problem along with a team of experts to help solve it. The possibility of controlling both the progress and some aspects of the outcome add dimension to the mystery that cannot be had with print or cinema alone. It is not coincidence that many other popular games use mystery as a device: mysteries are popular because humans are curious. One of the advantages that technology offers lies in its ability to prevent players from 'peeking'. In this format we can control very tightly the experience received by players – we can make them work through various parts while still giving the illusion of player control. However, the story itself must be presented in a way that entices players to continue. That is key.

Simply being led through a series of predetermined steps without any other thing to compel the participant will not sustain a voluntary learner, and all players of games are voluntary learners. A counter-example that comes to mind is an otherwise stunning online tutorial that quite literally leads learners through a virtual dissection of a frog called, *Froguts* (Hill & Hughes, 2001). The concept is promising, the facts are accurate, the animation is smooth and the imagery is excellent, but the interaction is so narrowly prescribed that after the first few operations we realize our role in this 'game' is simply to perform the next task. The application is in many ways quite engaging – the author happily worked through the entire Owl Scat demo to reconstruct the skeleton of the vole – but to call it a game is misleading as there is very little that is game-like about it, and to call it a simulation does a similar disservice to simulations everywhere. There is no way for us to explore any 'what if I do this' questions, which, by the way was the first thing this author tried when exploring the demo. How many children would have tried exactly the same thing? When I saw the quality of the graphics and the smoothness of the interaction, I immediately started to think of things to try that I would never have done in a real dissection but that might be fun to try in an environment where smell was not a factor, where there was no chance I might cut or poke myself, and where I would not be saddled with cleaning up the mess. I wanted to look inside its limbs, to try alternate organ arrangements or to see inside its head (we never got to do THAT with the real frogs in school). Instead, when we first begin, we are shown a frog laid out on a tray and ready, and we are given a box of pins to be used to secure the frog for dissection. We are shown four locations (marked by red 'x's) that are presumably the places we should put our pins. It is possible to put the pins anywhere, but once we have placed four pins, we cannot take any more out of the box, even though the box is not empty. Further, we can not proceed to the next step until we have placed the pins in the designated right locations. From there, our options for exploration diminish, and once we get to the point where we are using the scalpel, we must 'cut' along the pre-determined lines and in the pre-determined order. We don't even have a choice over where the scalpel is placed – all we can do is move it along its pre-determined path. We cease to be the 'scientists' and simply become the computer equivalent of a page-turner. This particular program is sold by annual subscription, and the 'home version' costs more than two *Phoenix* games. The frog dissection is advertised to be usable for one year, and *Phoenix* can be used until the cartridge breaks. The frog dissection is an award-winning program, yet still does not compare well when seen next to *Phoenix*. It is not a game and primarily calls itself a simulation. However, simulation need not be synonymous with animated film-strip, and this one unfortunately implements just a subset of what is otherwise a high fidelity simulation. I wonder if the folks at NASA would find their shuttle simulators as valuable a learning tool if the astronauts could only make the 'right moves' and never crash?

Animal Crossing is not an educational game yet still employs the same kinds of mechanisms we value in instructional design. Even though what is learned may not be valued in society, the players of this game nonetheless learn a large variety of skills, facts, principles and relationships. The game's content is frivolous, but its format is successful. Imagine what could be done with a format like this if our goal were an increased awareness and understanding of environmental issues? How about "*Wetlands*", or "*Rain Forest*"? We could make use of a format like this to help learners explore friendships or collaboration. An important element of both this game and *Phoenix Wright* is that of humor, and it is one of the motivators for exploring options in the game that we already know are unlikely to help us towards the overall game goal (Dormann & Biddle, 2006). A *Wetlands* game that was not visually appealing, or that never made us smile would not hold our voluntary interest for long, and part of the power of digital games is their ability to motivate and engage us.

Last but not least, Mario is certainly not a game that is likely to be seen along side games like *Civilization*, and *Zoo Tycoon* on any list of COTS (commercial off the shelf) learning games. Still, there is much that can be learned by studying the way it does what it does. *Mario* can keep people engaged for hours, while its educational counterpart, *Math Blaster* can not. *Mario* gives players constant and regular feedback so players always know where they are and where they need to go, and *Math Blaster* is a mystery tour. *Mario* allows players to back track and re-do completed challenges to gains points, lives, and power-ups; not only does *Math Blaster* offer none of these, the scoring mechanism is tied to the learning objectives of the math facts we are supposed to be practicing and we are given no clue as to how our score maps onto our in-game performance. Winning and loosing is tied directly to the equations we encounter while other game elements have little effect. The equivalent scenario in *Mario* would be if the red coins turned out to be what freed the Princess and defeating Bowser was coincidental. To someone focused primarily on formal learning outcomes these differences may seem trivial, yet they are the very things that compel players to remain engaged with a game for hours or discard it for something else after just a few minutes. Ignoring them is to waste some of the very qualities that make games and game technology useful.

5.10 Conclusion

I am always doing what I cannot do yet, in order to learn how to do it.

- Vincent Van Gogh

Although many references have been included in this chapter, most of the connections drawn between game elements, specific games, and the learning and teaching theories and models are being presented here for the first time. It is true that they are not really based on any prior work other than that of the author. Studies that examine the educational value of games have, for the most part focused on the player and what they get out of it rather than the game itself (Ellis, Heppell, Kirriemuir et al, 2006; Freitas, 2007; Mitchell & Savill-Smith, 2004). It is hoped that this work serves as the start of a conversation rather than the conclusion. It is not meant to be definitive, and will hopefully prompt others to help to clarify the game taxonomy as well as the mappings. The study of game design for educative purposes is still in its infancy.

Learning in games is highly experiential and situated. In that respect they are little different from the kinds of real-life contexts that are currently favored among scholars of instructional design and methodology. As we have seen, many games, while compelling and attractive can also bear little resemblance to anything we encounter in the real world. This does not mean that the format has limited applicability. What it does mean is that we should approach the design of educational games carefully with a thorough understanding of its potentials as well as its limitations. Just as this chapter has endeavored to connect the dots between accepted pedagogy and existing successful games, educator who use these devices must connect the dots between what is learned in the games and how that can be made authentic and applicable in the real world. After all, every learner deserves an answer to the questions, "Why am I doing this?" and "What is it good for?" even if they are being asked to play a game. Perhaps especially so in games.

5.11 References

Ausubel, D. P., Hanesian, H., & Novak, J. D. (1978). *Educational psychology: a cognitive view* (2d ed.). New York: Holt, Rinehart and Winston.
Becker, K. (2005). *How Are Games Educational? Learning Theories Embodied in Games.* Paper presented at the DiGRA 2005 2nd International Conference, "Changing Views: Worlds in Play", Vancouver, B.C., June 16-20, 2005.
Becker, K. (2006a). *Classifying Learning Objectives in Commercial Video Games: Proof of Concept.* Paper presented at the Canadian Games Studies Association Symposium, York, University, Toronto, Ontario, Sept 21-24.

Becker, K. (2006b). Pedagogy in Commercial Video Games. In D. Gibson, C. Aldrich & M. Prensky (Eds.), *Games and Simulations in Online Learning: Research and Development Frameworks*: Idea Group Inc.

Bloom, B. S. (1964). *Taxonomy of educational objectives; the classification of educational goals, by a committee of college and university examiners*. New York: D. McKay.

A Force More Powerful: The Game of Nonviolent Strategy BreakAway Games Ltd. (Designer) [Game] BreakAway Games Ltd. (Developer) (2006) [Windows] Published by International Center on Nonviolent Conflict & York Zimmerman Inc.

Bruner, J. S. (1966). *Toward a theory of instruction*. Cambridge, Mass.: Harvard University Press.

Phoenix Wright: Ace Attorney Capcom Co. Ltd. (Designer) [Game] Capcom Co. Ltd. (Developer) (2005) [Nintendo DS] Published by Capcom Co. Ltd.

Carolipio, R. (2006, 6/22/2006 12:00 AM). Playing with purpose: Video games are tackling serious issues like never before. *San Bernardino County Sun*

Math Blaster! Davidson, J. (Designer) [Game] Knowledge Adventure Inc. (Developer) (1986) [PC Game] [PC] Published by Knowledge Adventure, Inc.

Dobson, J. (2006). Survey: 'Word Of Mouth' Most Important For Game Buyers [Electronic Version]. *Gamasutra*, 2006. Retrieved Nov. 14 2006 from http://www.mi6conference.com/Magid_MI6.pdf.

Dobson, M., Ha, D., Mulligan, D., & Ciavarro, C. (2005). *From real-world data to game world experience: Social analysis methods for developing plausible & engaging learning games*. Paper presented at the DiGRA 2005 2nd International Conference, "Changing Views: Worlds in Play", Vancouver, B.C., June 16-20, 2005.

Dormann, C., & Biddle, R. (2006). Humour in game-based learning. *Learning, Media & Technology, Special Issue: Digital Games and Learning, 31*(4), 411-424.

Duffy, T. M., Lowyck, J., & Jonassen, D. H. (1993). *Designing environments for constructive learning*. Berlin; New York: Springer-Verlag.

Eagleson, A. (2006). Animal Crossing Wild World Game Walkthrough [Electronic Version]. Retrieved December 12, 2006 from http://db.gamefaqs.com/portable/ds/file/animal_crossing_ww_d.txt.

Real Lives EducationalSimulations (Designer) [Game] Educational Simulations (Developer) (2002) [Computer Game] [Windows] Published by Educational Simulations.

Egenfeldt-Nielsen, S. (2005). *Beyond Edutainment: Exploring the Educational Potential of Computer Games*. Unpublished PhD, IT University Copenhagen, Copenhagen.

Ellis, H., Heppell, S., Kirriemuir, J., Krotoski, A., & McFarlane, A. (2006). Unlimited Learning: The role of computer and video games in the learning landscape [Electronic Version]. Retrieved Dec. 10, 2006 from http://www.elspa.com/assets/files/u/unlimitedlearningtheroleofcomputerandvideogamesint_344.pdf.

Federation of American Scientists. (2006, Oct 25 2005). *Report on The Summit on Educational Games*. Washington, DC: Federation of American Scientists from http://www.fas.org/gamesummit/.

Freitas, S. d. (2007, *Learning in Immersive Worlds: a review of game based learning*. London: Joint Information Systems Committee (JISC) from http://www.jisc.ac.uk/media/documents/programmes/elearning_innovation/gaming%20report_v3.3.pdf.

Gagné, R. M. (1977). *The conditions of learning* (3d ed.). New York: Holt, Rinehart and Winston.

Gagné, R. M. (1985). *The conditions of learning and theory of instruction* (4th ed.). New York: Holt, Rinehart and Winston.

Gagné, R. M., Briggs, L. J., & Wager, W. W. (1992). *Principles of instructional design* (4th ed.). Fort Worth, Tex.: Harcourt Brace Jovanovich College Publishers.

Hadziomerovic, A., & Biddle, R. (2006). *Tracking Engagement in a Role Play Game.* Paper presented at the Future Play, The International Conference on the Future of Game Design and Technology, The University of Western Ontario, London, Ontario, Canada, October 10-12 2006.

Hill, R., & Hughes, D. (2001). Froguts. Retrieved Jan 12, 2006, from http://www.froguts.com/flash_content/index.html

Hopson, J. (2006). We're Not Listening: An Open Letter to Academic Game Researchers [Electronic Version]. *Gamasutra.* Retrieved Nov. 10, 2006 from http://gamasutra.com/features/20061110/hopson_01.shtml.

Jonassen, D. H. Design of Constructivist Learning Environments (CLEs). from http://tiger.coe.missouri.edu/~jonassen/courses/CLE/

Jonassen, D. H. (2004). *Learning to solve problems: an instructional design guide.* San Francisco, CA: Pfeiffer.

Jonassen, D. H., & Rohrer-Murphy, L. (1999). Activity theory as a framework for designing constructivist learning environments. *Educational Technology Research and Development, 47*(1), 61-79.

Juul, J. (2005). *Half-real: video games between real rules and fictional worlds.* Cambridge, Mass.: MIT Press.

Kücklich, J. (2005). Precarious Playbour: Modders and the Digital Games Industry. *Fibreculture Journal*(5).

Kuutti, K. (1996). Activity theory as a potential framework for human computer interaction research. In B. A. Nardi (Ed.), *Context and consciousness: activity theory and human-computer interaction* (pp. 17-44). Cambridge, MA: The MIT Press.

Lash, C. (2006). West Virginia schools use dance video game in gym class [Electronic Version]. *Pittsburgh Post-Gazette.* Retrieved Nov. 26 2006 from http://www.post-gazette.com/pg/06155/695356-298.stm.

Lave, J., & Wenger, E. (1991). *Situated learning: legitimate peripheral participation.* Cambridge [England]; New York: Cambridge University Press.

Leont'ev, A. N. (1978). *Activity, consciousness, and personality.* Englewood Cliffs, N.J.: Prentice-Hall.

Lurïëiia, A. R. (1976). *Cognitive development, its cultural and social foundations.* Cambridge, Mass.: Harvard University Press.

Merrill, M. D. (1999). Component Display Theory. In C. M. Reigeluth (Ed.), *Instructional-design theories and models* (Vol. 1, pp. 279-333). Hillsdale, N.J.: Erlbaum.

Merrill, M. D. (2002). First Principles of Instruction. *Educational technology research and development : ETR & D, 50 Part 3*, 43-60.

Merrill, M. D., Li, Z., & Jones, M. K. (1991a). Instructional transaction theory: an introduction. *Educational Technology, 31*(6), 7-12.

Merrill, M. D., Li, Z., & Jones, M. K. (1991b). Second generation instructional design. *Educational Technology, 30*(2), 7-14.

Minton, J. (2006, Jun 3, 2006). Video games seized from teen's home. 2theAdvocate.com. Retrieved Jun 15 2006, from http://www.2theadvocate.com/news/police/2924321.html?showAll=y

Mitchell, A., & Savill-Smith, C. (2004). The Use of Computer and Video Games for Learning [Electronic Version], 2004 from http://www.lsda.org.uk/files/pdf/1529.pdf.

Super Mario Bros. Miyamoto, S. (Designer) [Game] Nintendo (Developer) (1985) [Console Game] [NES] Published by Nintendo.

Super Mario Bros. 3 Miyamoto, S. (Designer) [Game] Nintendo Co. Ltd (Developer) (1988) [NES] Published.

New Super Mario Bros. Miyamoto, S. (Designer) [Game] Nintendo (Developer) (2006) [Handheld Game] [Nintendo DS] Published by Nintendo.

Making History: The Calm and The Storm Muzzy Lane Software (Designer) [Game] Muzzy Lane Software (Developer) (2006) [Computer Game] [Windows] Published by Muzzy Lane Software.
Animal Crossing Nintendo (Designer) [Game] Nintendo EAD (Developer) (2001) [GameCube] Published by Nintendo of America Inc.
Animal Crossing Wild World Nintendo (Designer) [Game] Nintendo EAD (Developer) (2005) [Nintendo DS] Published by Nintendo of America Inc.
Big Brain Academy Nintendo Co. Ltd. (Designer) [Game] Nintendo Co. Ltd. (Developer) (2006) [Nintendo DS] Published by Nintendo of America Inc.
Oliver, M., & Pelletier, C. (2004). *Activity theory and learning from digital games: implications for game design.* Paper presented at the Digital Generations: Children, young people and new media, London, July.
Pelletier, C., & Oliver, M. (2006). Learning to play in digital games. *Learning, Media & Technology, Special Issue: Digital Games and Learning, 31*(4), 329-342.
Prensky, M. (2001). *Digital game-based learning.* New York: McGraw-Hill.
Reigeluth, C. M., & Stein, F. S. (1983). The Elaboration Theory of Instruction. In C. M. Reigeluth (Ed.), *Instructional-Design Theories and Models: An Overview of Their Current Status* (Vol. 1, pp. 335-381). Hillsdale, N.J.: Erlbaum.
Reitman, J. (2003, 2003/11/13/). 'Fat boy'. *Rolling Stone,* 56-61.
Salen, K., & Zimmerman, E. (2004). *Rules of play: game design fundamentals.* Cambridge, Mass.: MIT Press.
Sallee, M. R. (2006). *New Super Mario Bros. Game Guide.* Retrieved Sept 13 2006.
Savin-Baden, M. (2000). *Problem-based Learning in Higher Education.* Buckingham: Open University Press.
Scardamalia, M., & Bereiter, C. (1994). Computer Support for Knowledge-building Communities. *The Journal of the Learning Sciences, 13*(3).
Schank, R. C., & Cleary, C. (1995). *Engines for education.* Hillsdale, N.J.: L. Erlbaum Associates.
Global Conflicts: Palestine Serious Games Interactive (Designer) [Game] (2007) [PC] Published by Serious Games Interactive.
Squire, K. (2002). Cultural Framing of Computer/Video Games. *Game Studies, 2*(1).
Sulpher, B. P. (2006). New Super Mario Bros. Game Walkthrough [Electronic Version], Version 1.7. Retrieved Feb. 12, 2006 from http://db.gamefaqs.com/portable/ds/file/new_super_mario_bros_d.txt.
Vygotsky, L. S., & Cole, M. (1977). *Mind in society: the development of higher psychological processes.* Cambridge Harvard University Press.

Chapter 6
Applying Pedagogy during Game Development to Enhance Game-Based Learning

Atsusi Hirumi and Christopher Stapleton

Abstract

"Serious games" are emerging as an important outgrowth of the video gaming industry. Entertaining games, such as Flight Simulator and SimCity, are already in use in schools and universities across the country, and the number of serious games designed specifically for training and education is also on the rise. Advances in video game production, however, are far outpacing research on its design and effectiveness. Relatively little is still known about methods for optimizing the game design process or game-based learning.

If educators and instructional designers lead the design process, the resulting game may be neither fun, nor engaging. Games that over-emphasize educational requirements often fall short of realizing the potential of play, game, and story for creating memorable experiences. Perceived learning requirements and traditional teaching practices may be forced onto the game, undermining the dramatic flow of story and disrupting the riveting interactions of gameplay. The game may be built on sound pedagogical foundations and incorporate proven educational practices, but if it is not fun or otherwise engaging, it will fail to meet the expectations of both producers and consumers. In contrast, if entertaining game designers dominate the design process, the game may fail to apply key pedagogical principles and players may leave entertained, but lack vital skills and knowledge. The importance and depth of content information and vital instructional events can be overlooked, oversimplified or trivialized while striving to uphold compelling goals of interactive entertainment. The game may distract players who may be enamored by the use of high-end graphics and animation, or by competing, scoring and winning, rather than learning.

This chapter posits a systematic process for designing serious games that integrates common instructional systems design (ISD) tasks with a game development process to optimize game-based learning.

6.1 Applying Pedagogy during Game Development to Enhance Game-Based Learning

Instructional [computer] games have reemerged as an important outgrowth of the video game entertainment industry. Video games are being repurposed for use in schools and universities across the country, and the number of original games designed specifically to facilitate training and education in conventional, hybrid and totally online courses is also on the rise. The problem is, like many rapidly growing industries, advances in video game technology are far outpacing research on its design and effectiveness. Relatively little is understood about how to apply what we know about teaching and learning to optimize game-based learning. For the most part, educators and instructional designers know little about game design and entertaining game developers may know little about education and instructional design. As a result, educators and instructional designers may not realize the potential of play, game, and story to create engaging and memorable learning experiences, and game developers may not apply basic pedagogical principles to optimize learning.

This chapter is written primarily for educators and instructional designers interested in game-based learning (GBL), to increase knowledge of the game development (GD) process and the capacity to communicate and collaborate with game developers. It is also written for game developers, to illustrate how basic instructional design (ID) tasks may facilitate learning and add value to the game design process. For all, we seek to highlight the value of collaboration between, as well as to stimulate dialog among professional educators, instructional designers and game developers to build stronger bridges between, and bring the seemingly disparate worlds of ID and GD closer.

Initially, five levels of application and key components of interactive entertainment are delineated to help educators determine the scope and purpose of instructional games and to set a context for the posited methods. Then, a basic GD process is characterized, including discussions of how educators can apply their knowledge of the subject matter, educational context, and pedagogy during the process to optimize game-based learning. Finally, the chapter concludes with a summary of key concepts and issues.

6.2 Levels of Design and Application

We posit that instructional games may be applied at four levels to optimize learning. As Driscoll (1994) suggests, instruction may be viewed as a series of events that are designed intentionally to facilitate learning and the achievement of specified learning and performance goal(s). Figure 6.1 illustrates how training or educational courses may consist of two or more instructional units. Instructional units, in turn, are said to contain a complete set of instructional events (or an instructional strategy) necessary to achieve a specified group of objectives. Units may also be divided into lessons, and for the purposes of this chapter, a lesson is said to consist of a subset of events related to a particular strategy.

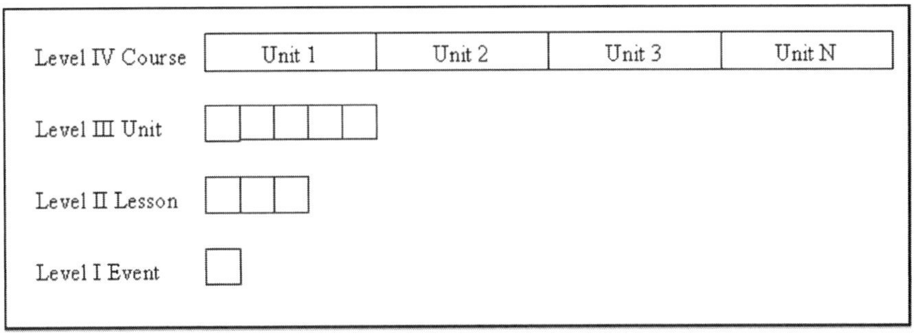

Level IV Course	Unit 1	Unit 2	Unit 3	Unit N
Level III Unit				
Level II Lesson				
Level I Event				

Fig. 6.1 Four Levels of Game Design and Application

Application of an instructional strategy "grounded" in research and theory further illustrates how games may be applied at alternative levels for different instructional purposes. For example, let's say a high school biology teacher decides to apply the Biological Sciences Curriculum Study (BSCS) 5E instructional model (BSCS, 2006) to facilitate his or her course as depicted in Table 6.1.

Table 6.1 Description of events associated with the BSCS 5E instructional model and adapted for the game environment

Engage	The instructor assesses the learners' prior knowledge and helps them become engaged in a new concept by reading a vignette, posing questions, doing a demonstration that has a non-intuitive result (a discrepant event), showing a video clip, or conducting some other short activity that promotes curiosity and elicits prior knowledge.
Explore	Learners work in collaborative teams to complete activities that help them use prior knowledge to generate ideas, explore questions and possibilities, and design and conduct a preliminary inquiry.
Explain	Learners should have an opportunity to explain their current understanding of the main concept. They may explain their understanding of the concept by making presentations, sharing ideas with one another, reviewing current scientific explanations and comparing these to their own understandings, and/or listening to an ex-

Elaborate	planation from the teacher that guides them toward a more in-depth understanding. Learners elaborate their understanding of the concept by conducting additional activities. They may revisit an earlier activity, project, or idea and build on it, or conduct an activity that requires an application of the concept. The focus in this stage is on adding breadth and depth to current understanding.
Evaluate	The evaluation phase helps both learners and instructors assess how well the learners understand the concept and whether they have met the learning outcomes. There should be opportunities for self assessment as well as formal assessment.

Adapted with permission from BSCS. The 5E Instructional Model is Copyright© 2008 by BSCS. All rights reserved.

At the *Event Level (I)*, a game may be designed to facilitate one specific instructional event within an instructional unit or lesson. An instructional game, for example, may be designed to facilitate recall of factual content or to promote active involvement and discussion (Dempsey, Lucassen, Haynes, & Casey, 1996; Blake and Goodman, 1999). In our example, a relatively simple "game-show" (e.g., jeopardy game) may be designed specifically to *evaluate* learners' acquisition of concepts and verbal information. The educator must then plan additional events prior to the use of the game, either online or face-to-face in hybrid learning environments, to *engage* learners, and to enable learners to *explore* the content, as well as *explain* and *elaborate* their findings. When applied as an instructional event, additional events and activities must often occur before and/or after gameplay to optimize learning.

At the *Lesson Level (II)*, a game may address two or more events contained in an instructional unit. For instance, a game may present learners with a scenario to *engage* their interest and ask them to *explore* related concepts through a series of readings and activities. Again, additional events, such as learner assessments and feedback, may have to occur before and/or after gameplay to facilitate learning.

At the *Unit Level (III)*, a game incorporates all of the events and activities necessary to achieve a specified set of goals and objectives associated with an instructional lesson or unit. In the BCSC example, that means the game will be designed to engage learners, facilitate exploration, solicit explanations and elaborations, and evaluate learning.

At the *Course Level (IV)*, one game is played throughout an entire course, tying together all the units, lessons and events associated with the course. It is conceivable that a game may be designed at a fifth, *Program Level (V)*, transcending all courses associated with a certificate, degree or training program, but the likelihood of such occurrences may be remote.

The distinctions are important because the process and resources necessary to apply what we know about teaching and learning (pedagogy) to design an instructional game may differ significantly depending on purpose and level of application. For instance, the process used to create a jeopardy game as an instructional event to evaluate learners' ability to recall verbal information may be relatively simple, involving the use of a free and easy to use jeopardy game shell and the preparation of questions and answers that are congruent with specified objectives.

In comparison, the process necessary to create an original game for an entire instructional unit or course may be much more complex. Game developers may have to write an original story, craft playful interactions, establish rules governing game play, and if a game engine is used to facilitate development, apply advanced programming skills, as well as generate graphics and animations. Although the concepts, process, and related resources discussed in the chapter may be applied to the development of Level I (event) and II (lesson) games, they are more commensurate with Level III (unit) and IV (course) games.

6.3 Fundamental Components of Interactive Entertainment

Each phase of the Game Design (GD) process offers developers an opportunity to enhance the learning experience. Educators are neither expected to participate in, nor contribute to, all aspects of GD. Rather, the challenge lies in determining when and how educators can apply their knowledge of pedagogy and the instructional situation to optimize game-based learning.

Stapleton and Hughes (2006) posit a framework that helps illustrate how the results of fundamental ID tasks may be used to facilitate the design of instructional games. The framework identifies three key components of interactive entertainment, each further delineated by three interrelated elements that must all work together to create memorial experiences. The components and elements elucidate key questions, noted in Table 6.2, that educators, instructional designers and game developers answer in increasing detail to integrate and balance requirements for learning and entertainment as the team progresses through the GD process.

Table 6.2 Three Basic Components and Related Elements of Interactive Entertainment

Components	Elements
Story WHY should I care, from player's point of view?	Characters – WHO is it about & who am I in that experience?
	Events – WHAT & WHEN do things happen to motivate me or propel action?
	Settings – WHERE am I and what does it matter (Context)?
Game HOW do things work (procedural or mechanics)?	Goals – WHAT for (Aspirations related to Why)?
	Rules – WHY not (Limitations and restriction to play, strategy, etc.)?
	Tools – With WHAT devices (special effects, instruments, etc.)?
Play WHAT am I doing (Participatory)?	Cause – IF I...(stimulus)?
	Effect – THEN the game... (response)?
	Consequences – THEN, I will... (consequences)?

By applying pedagogy and knowledge of the subject matter, learners, and instructional context to answer the questions, designers flesh out the core game plan and reconcile game and learning goals so that the entertainment supports the learning and the learning enhances the entertainment. The more the learning content and objectives are interwoven into the entertainment elements, the more the game will reinforce the learning objectives. The framework, along with its basic components, elements, and questions, will be referred to throughout the chapter to discuss how educators may contribute to the design and development of an instructional game.

6.4 Applying Pedagogy during the Game Development Process

Both instructional design (ID) and video game development (GD) processes consist of comparable phases. Table 6.3 identifies key tasks associated with each phase of the ID process and relates them to the phases and specific [interim] products generated during the GD process.

Table 6.3 Relating Instructional Design Tasks with Game Development Products

ID Process and Tasks	GD Process and Products
Analysis Phase	*Concept Development Phase*
• Assess needs and identify goal(s) • Analyze goal(s), learner and context	• Prepare pitch document • Prepare game concept document
Design Phase	*Pre-Production Phase*
• Generate, cluster & sequence objectives • Determine learner assessment method • Generate instructional strategy • Select media	• Create game design documents • Prepare art bible and production plan • Create technical design document
Development Phase	*Prototype & Production Phases*
• Acquire materials or outsource development • Create flowcharts and storyboards • Generate prototypes • Formatively evaluate and revise materials	• Develop analog or low-fidelity prototypes • Develop tangible prototypes • Produce Alpha Version • Produce Beta Version • Produce Gold Version
Implementation & Evaluation Phases	*Post-Production*
• Deliver and manage instruction • Plan and conduct summative evaluations	• Generate and release subsequent versions • Generate and release upgrades/expansions

The specific tools, tasks and techniques used during each phase may vary by project and by organization. Organizations may also follow a waterfall or spiral approach when applying the process, but the phases remain basically the same. The ID process typically consists of analysis, design, development, implementation and evaluation, and GD process often consists of concept development, pre-production, production, and post-production. An understanding of the relationship between the ID and GD processes will help instructional designers and educators collaborate with game developers and visa versa. It also helps identify when and what instructional designers and educators can do to apply their knowledge pedagogy and the subject matter. The following is a brief summary of what game developers typically do, along with a more detailed discussion of what instructional designers and educators can do during each phase to facilitate game development and optimize game-based learning.

6.4.1 The Concept Development Phase

The concept development phase begins when a game is first conceived and ends when a decision is made to fund or otherwise support the development of the game and initiate additional planning. The goals are to determine what the game is about and convey key ideas to key decision makers and potential funding agents in a clear, concise written form. Initially, developers compare and contrast benchmarks games and related instructional programs (if they exist) to formulate and brainstorm ideas. They also may create crude paper prototypes and stage live improvisations of alternative game concepts, allowing them to flesh out mediocre ideas, expose core misconceptions and assumptions, and spark discovery of truly original ideas. Focus groups may then be used to verify audience's expectations and validate the entertainment value and flow of selected ideas before a game proposal and concept document are finalized.

In general, a concept document includes short descriptions of:

- The Premise or High Concept (describing the basic idea or "hook" that will make the game exciting and sets it apart from other games),
- Player Motivation (e.g., the game's victor condition),
- Game Play (what the player will do while playing the game),
- Story (main events, characters, and settings),
- Target Audience/Market,
- Game Genre,
- Target Platform and Hardware Requirements,
- Competitive Analysis, and
- Game Goals.

A brief game proposal (also referred to as a "pitch document") addresses the same topics as the concept document but in an abbreviated form, and is used during meetings, along with a "promotype" (a promotional demonstration of a key portion of the best scenes), to pitch or sell the game to potential supporters. If given the green light (and funding) to continue planning, developers enter the Design or Pre-Production Phase.

Instructional designers and educators can play a key role in, and help game developers formulate ideas, create a "promotype," and prepare persuasive pitch and concept documents by delineating and clearly communicating (a) targeted learning goals and related skills and knowledge, (b) key learner characteristics, (c) important contextual factors, and (d) a desired instructional approach.

Define Learning Goals and Subordinate Skills and Knowledge. To formulate preliminary ideas, game developers must have a good understanding of what learners are expected to know and be able to do as a result of the game. Specifically, knowledge of targeted learning goal(s), and the subordinate skills and knowledge necessary to achieve the goal(s), will help developers begin to answer the questions, "Why should I care?" and "What for?" as posed in Table 2. A list of targeted skills and knowledge is also necessary to identify, compare and contrast, and find correlations between successful and unsuccessful benchmarks of similar games and instruction.

Professional organizations and accrediting agencies may list skills and knowledge to be addressed by the instruction. Relevant learning goals and objectives may also be found in existing course materials and syllabi. In such cases, educators may review the list and select skills and knowledge to be addressed by the game, keeping in mind the availability of resources, level of application, and the scope of the proposed initiative.

In other situations, a list of goals, skills and knowledge may not exist. Educators may need to complete a goal, task, subordinate skills, subject matter or other forms of analyses (c.f., Jonassen, Tessmer, & Hannum, 1999) to identify relevant skills and knowledge. The analysis may also identify required entry behaviors (pre-requisite skills and knowledge learners must have to successfully initiate and complete the game).

Knowledge of the target learning goals, and the subordinate skills and knowledge necessary to achieve the goal help answer questions related to the story and game components of interactive entertainment posed in Table 6.2 at the highest and simplest level. Furthermore, to support game development, potential adopters, sponsors and funding agents will want to know what specific skills and knowledge are going to be addressed by the game to ensure its relevance and see how the game may fit with the larger curriculum. Thus, game developers should communicate learning goal(s) along with the entertainment goal(s) within the concept document.

Characterize Learners. Game developers recognize that the more you know about your customer, the more products you can sell them. In developing Walt Disney Imagineer's Ten Commandments, Sklar (n.d.) established the first two commandments of interactive entertainment as "know your audience" and "walk in your guests shoes." To formulate game ideas that match what's in the audience's mind, game developers analyze key psychographic variables (e.g., values, attitudes, lifestyles) as well as the basic demographic (e.g., gender, age, generation) of the target market through available research and market critiques. Learner analysis is the equivalent practice within ID.

Learner analysis reveals key characteristics of the target audience including: (a) entry behaviors, (b) prior knowledge of topic, (c) attitudes toward content and delivery system, (d) academic motivation, (e) educational and ability levels, (f) learning preferences, (g) attitudes toward the organization giving the instruction, and (h) group demographics (Dick, Carey, & Carey, 2005). During concept development, the integration of learner and target market analyses is posited to further delineate key learner characteristics and clarify learner preferences and expectations.

One of the keys to entertainment is to "satisfy expectations." An analysis of learners' attitudes toward the delivery system, content and organization reveals learner expectations that may then be used to hook the audience by creating empathetic and archetypal game characters with similar hopes, dreams and fears of the target audience. Knowledge of learners' general academic motivation and educational abilities may also help developers create challenges that are neither too tedious, nor too frustrating. A critical aspect of game design is to provide challenges that are hard enough to be demanding, but not annoying, while easy enough to be interesting, but not boring. Since the line between frustration and boredom may differ with every user, data on learners' academic motivation and educational abilities may be particularly useful for designing game challenges, as well as creating the algorithms necessary to monitor the behavior of the user to adapt the level of difficulty during game play. An analysis of key learner characteristics establishes the target audience's needs and wants, helping developers define the expectations that the game will need to set-up and satisfy.

If educators have sufficient knowledge of the target learner population to accurately portray key learner characteristics and help game developers address key game issues as discussed above, then they should document and communicate their knowledge of learners to game developers at the beginning of the concept development phase. If educators do not have adequate knowledge of the targeted learners, they should then consider conducting a learner analysis before or at the onset of concept development to help developers formulate game ideas that will address learners' needs, interests and expectations. Learner analysis results should also be integrated with the description of the target market within the concept document to demonstrate and help convince potential backers that the team has sufficient knowledge of the target learner population to develop a successful instructional game.

Characterize Learning and Performance Context. During concept development, knowledge of both the learning context (where learners are expected to acquire targeted skills and knowledge) and performance context (where learners apply newly acquired skills and knowledge) can also help developers answer key questions related to play, game and story, noted in Table 6.2.

Knowledge of the physical, social and psychological aspects of the performance context is essential for developers to generate appropriate mock ups of the game world, and provides valuable clues on how to excite players with key character interactions and game play. Facts about the physical conditions under which learners are expected to demonstrate targeted skills and knowledge, including the availability, use and nature of facilities, equipment and other resources, are vital for game developers to create appropriate settings within the game. Awareness of the social networks that form in regards to targeted learning and performance outcomes is also useful for designing player and non-player character interactions, particularly if the game is to simulate workplace relationships. Psychological conditions, such as the degree and nature of stress typically felt as learners perform related job tasks may also provide rich enhancements to increase immersion.

Knowledge of the learning context is also necessary to define basic technical requirements reported in concept documents. In many educational situations, the learning environment may differ from the performance setting where newly acquired skills and knowledge are to be applied. For example, the learning environment for a totally online course may consist of networked computers with access to a learning management system and the instructional game. In comparison, a hybrid learning environment may include a networked computer, as well as periodic face-to-face meetings at school. Knowledge of key variables, such as the configuration and connectivity of computers used at home and school, can help define hardware requirements for the concept document and to ensure the game will work with available equipment and resources.

If educators do not have extensive knowledge of the performance or learning setting, they should consider completing a context analysis, as prescribed by ID professionals (e.g., Dick, Carey, & Carey, 2005; Smith & Regan, 2005). The analysis provides vital information about the performance setting, such as (a) managerial/supervisory support, (b) physical aspects of the site, (b) social aspects of the site, and (d) relevance of skills to workplace. The analysis also reveals important information about the learning context, including, but not limited to the (a) number and nature of sites, (b) compatibility with instructional needs, (c) compatibility with learner needs, and (d) feasibility for simulating the workplace. Information about the context should be conveyed early in the concept development phase to help game developers answer the key questions related to story, game and play (noted in Table 2) at the simplest level. Context analysis results should also be documented within the concept document to help garner support for game development.

Select Basic Instructional Approach. Fundamentally, instructional games differ from entertaining games in that they are designed intentionally to facilitate achievement of specified learning goals and objectives. We argue that the application of pedagogy is necessary to facilitate achievement and optimize game-based learning. During concept development, the selection of a basic instructional approach provides valuable insights into how content information is to be presented to learners and the nature of interactions that are designed to facilitate game-based learning. Communication of the instructional approach may also help garner support for continued design and development.

Will the game be based on behavioral, cognitive information processing, constructivist learning, or brain-based learning principles? Should it apply a specific instructional strategy, model or theory, such as Case-Based Reasoning (Aamodt & Plaza, 1994), Learning by Doing (Schank, Berman & Macpherson, 1999), or Problem-Based Learning (Barrows, 1985)? The selection, and later application, of a basic instructional approach is critical for determining the nature of the learning environment and guiding the overall design and sequencing of critical learning interactions and game play that ultimately affects the manner in which learners achieve specified learning outcomes.

For example, if a behavioral learning approach is selected for a game, the "story" may present learners with key facts, concepts and/or principles and the "game play" may ask learners to respond to discriminative stimuli, such as a question. Based on learners' responses, the game will then present learners with contingent stimuli that serve to either reinforce or discourage similar responses. Many drill and practice games, popular in the 1980s and 90s were based on behavioral learning principles.

In contrast, if a constructivist approach is selected, the "story" may present learners with a scenario or problem and the "game play" may require learners to utilize various tools to access content information, derive meaning, and construct their own knowledge of how to work their way through the scenario and/or solve the problem. Moreover, if a social constructivist approach is selected, then a multi-player game may be conceived that emphasizes the use of tools for player-player interactions designed to facilitate the social construction of knowledge. Although details on how the game will apply key principles, tools and events associated with a particular instructional approach are addressed during pre-production, the selection of a basic instructional approach is critical in defining the broader interactions of the learning experience and informing high-level interactive entertainment design. By basing the early entertainment development on pedagogy, any subsequent artistic choices will most always enhance, rather than obstruct achievement of the learning objectives.

The selection of an instructional approach requires developers to consider the nature of the desired learning outcomes, his or her personal beliefs about teaching and learning, as well as the values of potential supporters and adopters. The nature of the desired learning outcomes should drive the design process. For example, to facilitate recall of verbal information, or to train people on a relatively simple procedure, a direct instructional approach based on behavioral learning principles (e.g., Joyce, Weil, & Showers, 1992) or information processing theories of learning (e.g., Gagne, 1977) may be more appropriate than constructivist or learner-centered methods. In contrast, if the desired learning outcome requires higher-order thinking, where there may be more than one correct answer or more than one method for deriving the correct answer, then constructivist (e.g., Jonassen, 1999; Wilson, 1995) or related experiential (e.g., Kolb, 1985; Egenfeldt-Nielsen, 2005), case-based (e.g., Leake, 2000; Aamodt & Plaza, 1994), or problem-based approaches (e.g., Barrows, 1985; Savery & Duffy, 1995) to teaching and learning may be favored.

In selecting an appropriate instructional approach, it is also important to take in account the educational philosophy and epistemological beliefs of developers' as well as potential adopters' and supporters. If the developers believe that people derive meaning and construct knowledge through social interactions, then the selection of constructivist, learner-centered, and cooperative instructional approaches may support his or her beliefs. In contrast, if developers believe people learn by processing information through sensory, short-term, and long-term memory, than an instructional approach based on information processing theories of learning may resonate with his or her educational philosophy (e.g., Gagne, 1977). If the designers are pragmatists and believe that meaning is constructed by individuals based on their interpretation and understanding of reality, she or he may prefer an eclectic approach, selecting from a range of behaviorist to constructivist instructional methods depending on the situation.

The educational values and beliefs of potential supporters and adopters should also be considered. If the pedagogical foundations of a game is not congruent with the beliefs and values of instructors, administrators, managers, or other potential supports and stakeholders, then the chances of the game being approved for development, or adopted for use, are relatively slim. Furthermore, if developers are working with an instructor and/or subject matter expert to create an instructional game, then it may be important to discuss and, if necessary, reconcile any differences in philosophy prior to initiating selecting or applying a basic instructional approach.

Selecting an appropriate instructional approach is neither simple, nor straightforward. Much depends on the desired learning goals and objectives, but concerns for the fundamental beliefs about teaching and learning also mediate the selection process. Whatever theory or approach is selected, related principles, events and interactions provide a rich resource for inspiring game developers, if presented in the terms they understand (e.g., concrete verbs, events, exchanges, rather than theoretical propositions). Game developers, in this early phase of conception, are looking for these types of sources to challenge the imagination and spark creative correlations between entertainment and education.

A basic instructional approach should also be communicated in the concept document to garner backing for the game initiative. Providing a short description of the pedagogical foundations may help convince potential supporters, particularly those with a strong educational background and/or investment, to back development. Holland, Jenkins, and Squire (2002) also believe that instructional games should be based on a pedagogical model to help ensure players are able to apply what she or he learned to real world contexts. A game is a tool not a magic box; game developers can ensure learning takes place only if it is well designed based on a solid pedagogical foundation. Whatever theory or approach is selected, game developers should clearly communicate and assure potential supporters that the instructional game will be based on a solid pedagogical foundation. The subsequent application of alterative instructional approaches is discussed further under Pre-Production.

6.4.2 Pre-Production Phase

After receiving support for the game concept, developers enter the planning or pre-production phase. It is during the pre-production phase that developers flesh out the details. What is the specific nature of the characters, events and settings that players will interact with during the game? How will they be presented to players? What rules and tools will govern and facilitate achievement of game goals? Analog prototypes may be created to help developers answer these, and other basic design questions and to ensure the game play mechanics are correct and the game is fun and compelling with no distractions resulting from visual style and programming features that are secondary to a game's foundation. Detailed game design and a technical design document are also prepared, along with art bible and a production plan, outlined below.

Although the organization and contents of game design documents differ by author, organization and game genre, they represent an extension of the concept document and typically include, but are not limited, a project overview and detail descriptions of:

- Story (characters, settings, and events)
- Game play (rules, tools and goals)
- User Controls
- User Interface
- Artificial Intelligence
- Game Levels
- Art, Audio and Technical Features
- Production Details
- Risk Analysis
- Development Budget

The Art Bible establishes the look and feel of the game and provides a reference for other art. It helps ensure consistency in style throughout the game and typically consists of:

- A set of visuals (ranging from pencil sketches to digitized images that capture final look of the game); and
- A visual reference library that reflects the direction the art should take over time.

The Technical Design Document is based on the Game Design Document and is typically written by the game's technical lead or director and includes a description of:

- The game engine, including comparisons with other engines on the market;
- How game will transition from concept to software;

- Who will be involved in the development of the game engine, including what tasks each person will perform, and how long it will take to perform each task; and,
- What core tools needed to build the game; including hardware and software that must be purchased.

The pre-production phase typically ends with the development of a tangible, digital prototype that represents, "...a working piece of software that captures on screen the essence of what makes the game special, what sets it apart form the rest, and what will make it successful" (Novak, 2005, p. 332). As Bates (2004) suggests, the prototype, "...can be the single greatest influence on whether the project goes forward. Publishers [and other funding agents] like to be able to look at a screen and 'get it' right away. If they can't see the vision within a minute or two, they're unlikely to fund the rest of the project" (p. 211). Where the promotional focus of the "promotype," created during concept development, may improvise solutions, it may also result in more questions than answers to feed key objectives of the design process. The "prototype," generated during pre-production, actually solves key problems with chosen tools to validate assumptions and provide a more detailed preview of the game than the "promotype."

To help developers create related design documents and answer the questions related to play, game and story, depicted in Table 6.2, in greater detail, educators should consider (a) generating, clustering and sequencing objectives, (b) delineating learner assessment methods, and (c) applying grounded instructional strategies and events. Educators should also consider conducting several formative evaluations to validate design and improve the prototype before it is presented to sponsors and other key stakeholders to demonstrate the developers' capacity to achieve specified goals, and garner continued support for production.

Generate, Cluster, and Sequence Objectives. During the Design Phase, learning objectives should be generated, clustered and sequenced to help game developers establish game levels, and define the overall story and game structure. Most (larger) games are broken up into levels or worlds. Levels divide a game into sections, organize progression, and enhance game play (Novak, 2005). A player usually needs to meet specific goals or perform specific tasks to advance to the next level. Often, levels may be similar but more difficult as a player progresses through the game. According to Novak, developers should consider game objectives, flow, duration, availability, relationships and difficulty when designing levels; in much the same way educators organize a course or training program into instructional units and lessons.

Without clearly defined objectives, players may randomly move, shoot, solve-problems or collect things to progress through the game. To help players focus their efforts, game developers communicate game objectives by creating a cut-scene or a short tutorial at the beginning of a level, or players may complete relatively simple tasks that illustrate the basic objectives of a game. Whatever method is used to communicate objectives, game developers believe that players should be informed of where they stand in relation to the overall game. Educators and instructional designers hold similar beliefs, recognizing the importance of focusing learners' efforts, and informing them of where they stand in relation to the overall course and learning goal(s). The specification of objectives may also be particularly important for instructional games. The ability to engage learners through fun and entertainment is one of the primary reasons why instructional games are being developed across settings. One concern facing potential game adopters is that learners may be distracted or may lose sight of desired learning outcomes. Concrete learning objectives may be presented to learners in the same fashion and time that game objectives are presented to players. This may help learners keep their focus of targeted learning outcomes.

In terms of flow, game developers may want players to stay in a particular area until they have accomplished certain objectives, or prevent players from returning to particular areas once they have completed specified objectives. The time, or duration, spent at each level may depend on the nature of the player. Novak (2005) refers to a "universal rule" that suggests a player should be able to complete at least one game level in a single session. For younger and/or novice gamers, one level may be designed to take 15 minutes; for older and/or advanced gamers, two hours of continuous concentration on a level may be acceptable. Educators also take into account flow, time, and the nature of learners when defining the scope and sequence of instructional units and lessons.

Availability refers to both the total number of levels included in a game, as well as the accessibility of levels within a game at any given time. The number and accessibility of levels depend on the nature of defined game goals and objectives. Each level should address one game goal or a logical set of objectives. In general, games should allow players to access as many levels as possible to maximize flexibility and address individual needs and interests, but accessibility must also be restricted to avoid confusion. Similar issues and concerns regarding availability must be addressed as educators and instructional designers work with game developers to cluster and sequence learning objectives and define game levels.

The relationship between levels must also be considered in terms of scenes, episodes or events within the larger story played out during the game. In some games, levels are defined based on story structure, where each level may be self-contained, with its own subplot, set of events and conclusion. For example, a strategy game may consist of a series of quests, campaigns or missions that need to be completed to finish the game. In other (puzzle) games, levels may be defined by increasing degrees of difficulty. The clustering and sequencing of learning objectives helps game developers define a suitable story structure and an overall experience arc for players as they progress through the game.

Figure 6.2 compares the structure of a course that may be divided into units, lessons and events, within the game, and organized into levels, scenes, and events, to form an overall experience arc (Stapleton & Hirumi, in press). The experience arc is designed to increase investment and emotional engagement and provide the overall satisfaction of achieving the ultimate goal, whether for entertainment or learning. The rewards and pay-offs need to propel the motivation of the player forward through the experience arc. The developers need to build up to the intended intensity by increasing risks and balancing the difficulty level between frustration and boredom so to offer appropriate challenges and game play. As developers define game levels, they must consider how the clustering and sequencing of instructional events serve the creative intent of the game as articulated by the experience arc. The learning goal should be interrelated with the overall game goal. Each unit should correspond to the emotional waypoints of each game level. Each lesson should integrate their instructional events with interactive exchanges of each scene of the game.

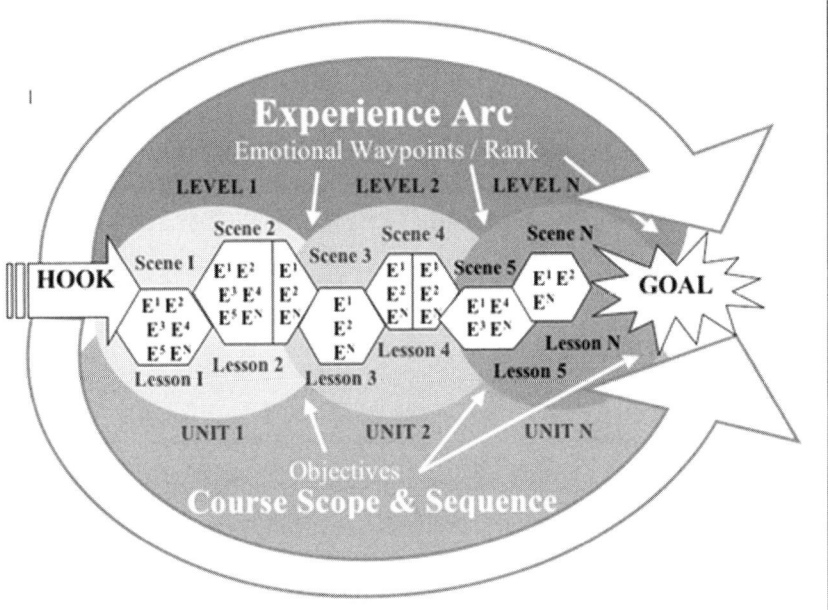

Fig. 6.2 The Similarities between the Organization of a Game and Instruction

The degree of difficulty is also an important consideration when defining game levels. As a player progresses through levels, the degree of difficulty may remain constant, increase linearly, or follow an s-curve, with the degree of difficulty remaining relatively flat at the beginning and end, but increasing exponentially during the middle. Novak (2005) also suggests that to challenge expert players, developers can build more difficult versions of each level that are accessible separately, or developers can build different degrees of challenge within each level. As educators and instructional designers cluster and sequence objectives, and work with game developers to define game levels, they must also consider the difficulty of specified objectives. Should the game present learners with objectives that increase in difficulty, remain constant or follow an s-curve over time? Should developers create additional levels and/or provide access to varying degrees of learning challenges with each level?

As noted under concept development, objectives may be derived from standards published by professional organizations, accrediting agencies or existing instructional materials. If objectives do not exist, educators may apply alternative analysis techniques to identify skills and knowledge necessary to achieve a specified learning goal, and the skills and knowledge may then be used to prepare objectives. At the onset of pre-production, educators should cluster and sequence objectives, and may use a simple course scope and sequence table, as depicted in Table 6.4, to communicate the organization of objectives to developers. Educators may then work with game developers to finalize the clustering and sequencing of objectives and to define game levels, by taking into consideration both game and instructional flow, duration, availability, relationships and difficulty. Educators should then use the objectives to delineate learner assessment methods that are congruent with the specified objectives.

Table 6.4 Template I for Basic Course Scope & Sequence Chart

Course [Learning or Performance] Goal Statement:			
Unit 1	Unit 2	Unit 3	Unit 4
Terminal Objective 1.0			
Enabling Objective 1.1			
Enabling Objective 1.2			
Enabling Objective 1.3			
Enabling Objective 1.N			

Delineate Learner Assessment Methods. Accurately assessing what players learned from an instructional game may be one of the greatest challenges facing game developers. It requires knowledge of what, when, where and how to assess learners' skills, knowledge, attitudes and abilities. The specification of measurable learning objectives delineates *what* needs to be assessed. Concrete entertainment goals and objectives (e.g., enhancing learner engagement, creating suspense, evoking emotion, promoting players' continuing motivation to return to similar goal directed behaviors) also help determine *what* needs to be assessed. The challenge lies in determining *when*, *where* and *how* to assess achievement of specified goals and objectives.

The fundamental ID tasks of determining learner assessment methods and aligning assessment with specified objectives provide further insights into assessment for game-based learning. Learner assessments may be completed before, during and after instruction. Before instruction, entry-level tests may be used to determine if learners have pre-requisite skills and knowledge and pre-tests may be given to determine if learners already have the skills and knowledge addressed by the instruction. During instruction, practice-tests may be used to help learners develop skills, acquire knowledge, and monitor their progress toward specified objectives. After instruction, post-tests measure learners' achievement of specified objectives. In an instructional game, assessments may be given: (a) before learners begin to play the game to determine if they have pre-requisite skills and knowledge necessary to successfully play and complete the game; (b) at the beginning of one or more game levels to determine if learners have the skills and knowledge addressed by the game level and adapt learners progress through the level accordingly; (c) during or at the end of one or more game levels to practice applying skills and knowledge and monitor learners progress toward specified objectives, or (d) after learners have completed the game to assess learner achievement of specified objectives.

To decide *when* to best assess learners' skills and knowledge, game developers must determine the importance, feasibility and cost of assessing whether learners: (a) have pre-requisite skills and knowledge; (b) have the skills and knowledge to be addressed by the instructional game; (c) need practice or help monitoring their acquisition and/or progress toward specified learning outcomes as they play the game; and (d) have acquired, can apply, and/or transfer targeted skills and knowledge to move up game levels or after completing the game. Answers to these questions may be derived from learner and context analyses (discussed earlier) and as developers generate an instructional strategy and integrate the strategy with the story and game play (as discussed in proceeding sections of the chapter). Determining *when* to assess learners' skills and knowledge also requires knowledge of *where* and *how* learners are to be assessed.

Fundamentally, game developers must decide *where* assessments are to take place, either within the game or outside of the game, which leads directly to the next assessment question; *how* to assess learners. If learner assessments are to be integrated within the game, game developers must have the skills and resources to formulate the algorithms (if/then statements) and create the artificial intelligence (AI) necessary to program the game to respond properly to learner input.

Assessment algorithms and AI may be relatively simple and programmed within a game if developers choose to use conventional criterion referenced testing methods (e.g., multiple-choice, true/false, matching, fill-in-the-blank) or product and performance checklists to assess learners when there is one correct answer, one correct method for deriving the answer, or one set of readily observable/recordable behaviors that demonstrate targeted skills and knowledge. However, if learner assessments require the application of a set of heuristics and some level of subjectivity to measure problem-solving and the use of higher order thinking skills within ill-structured domains, the algorithms and AI may be too complex for the team to formulate and apply, and the assessment may have to occur outside of the game with experts assessing the achievement of related objectives by examining learner generated work samples. Because the descriptors tend to be too imprecise for a computer to match specified criteria to students' behavior (Leddo, 1996), "performance assessment using scoring rubrics...that describe different levels of proficiency is considered unsuitable for computer games..." (Mitchell & Savill-Smith, p. 50).

During pre-production, game developers should determine learner assessment methods immediately after defining, organizing and classifying learning objectives to help ensure alignment between objectives and assessments. Too often, assessments are defined in isolation, with little to no consideration of the objectives. As a result, learners are often left wondering about the origins of specific test items or assessment criteria. To determine appropriate learner assessment methods, ensure assessment items and criteria are aligned to specified objectives, and communicate assessment methods to developers, educators may choose to complete a learner assessment alignment, as depicted in Table 6.5.

Table 6.5 Sample Learner Assessment Alignment Table

Skill	Objective	Domain	Method	Item/Criteria
Self-assess prior knowledge	1.0 Given a systematic design process, assess your prior skills, knowledge, interests and experiences relative to course topics and tools.	Cognitive Strategy	Post-Test: Portfolio Assessment Rubric	Proficient Performance Criteria: • Includes job title and teaching/training experiences. • Assesses prior knowledge of training and instruction. • Describes expectations and desired outcomes. • Communicates information in concise manner with few errors. • Posted to correct location by specified deadline.
Identify benefits	1.1 Given an instructional situation, you will be able to identify benefits associated with applying systematic design tools and techniques.	Verbal Information	Practice-Test: Conventional Multiple Choice	One of the primary benefits of systematic design is that it: (a) does not take too much time or resources. (b) is dynamic and helps adjust for learners' needs as they change during the instructional process. (c) ensures the alignment of test items with instructional objectives and strategies. (d) focuses on technology-based instruction. (e) all of the above.

In Table 6.5, column one lists essential skills and knowledge as identified by a goal, subordinate skills or other forms of analysis (as discussed during concept development). Column two represents the corresponding objective statement. Column three classifies the objective according to a learning taxonomy (in the example, Gagne's taxonomy of learning outcomes) to guide selection of an appropriate assessment method. Column four notes when the assessment should take place (pre, practice or post), and the prescribed assessment format (e.g., conventional multiple choice, true/false, matching, fill-in-the-blank; product or performance checklists; or assessment rubrics). Column five specifies the assessment criteria or item that should be used to measure achievement of the specified objectives.

Educators can use the table to ensure alignment between assessments and objectives by making sure the behavior required to successfully meet the assessment criteria or complete the assessment item is congruent with the behavior specified in the corresponding objective. The table also communicates when and how assessments may be used to facilitate and measure achievement of specified objectives. Game developers, in turn, may use the information to determine if the team has the capacity to develop the algorithms and AI necessary to apply the assessment items or criteria, or if assessments may have to occur outside of the game environment. If developers choose to integrate assessments within the game, the challenge again lies in the creative ability of developers to reconcile differences in entertainment and education, in this case, by making failure fun and creating assessments that are engaging, valid and reliable.

The assessment of learner performance provides the opportunity to deliver the culminating pay-off for both the game and instruction design. When the learner is in the height of immersion, making decisions, taking risk, and instigating action, they are forming memorable experiences and more likely to remember the lessons learned. If differences in education and entertainment are not resolved, resulting assessment will compete and dilute the intension of both by distraction or disassociation. For instance, let's say we were using a first-person-shooter game genre to teach grammar. For the level assessing the use of homonyms, a multiple choice activity was the appropriate and recommended assessment method, but if learners had to stop play to use a drop down menu to decide which homonym to use, it would break the flow (emotional and physical momentum), be anti-climatic and kill both the game objective and instructional appeal. However, time, accuracy and selection are integral to both multiple-choice and shoot-don't-shoot scenarios. The developers' desire and job is to make choosing fun by applying a form of play that is popular and proven. In this case, a first-person-shooter genre provides the type of game mechanics that is congruent with the recommended assessment method. The theme can then adapt to the learner's demographics to determine whether you use a machine gun to shoot the colonel (or kernel) or a magical wand to select a 24 karat (or carrot) treasure. Both the correct and incorrect answer can have entertaining and relevant consequences.

When the learning objectives are integral to the game objectives, the assessments can become the culminating experience where performance determines scoring and rank that provides extra powers and resources to take on to the next level. However, the assessment must be designed as creatively as the game play where making mistakes are as memorable as succeeding. Making losing fun provides not only the incentive to try again, but you form distinct memories of what not to do. The common mistake is to make the assessment an after thought or making it merely a dressed up multiple-choice question.

Apply Grounded Strategies and Events. Story and instruction may both be viewed as the deliberate arrangement of events. Story events, however, are written primarily to entertain and engage the audience. In comparison, instructional events are designed principally to facilitate learning and the achievement of specified learning objectives. We posit that grounded instructional strategies and events should be integrated with story events to optimize game-based learning. During preproduction, developers reconcile similarities and differences in instructional and story events to design meaningful interactions that facilitate learning in a fun and engaging manner.

Stories evoke emotional investment in game play. It answers the question, "why should I care?" The author develops characters that the audience can identify with to invite them into the action, and once empathic to the characters' plight, the audience goes wherever they are lead. Story events become consequence generators that drive future action and the setting provides the context of the action to reveal meaning. Once the story hooks the audience emotionally, the author elicits active participation by designing meaningful interactions between story events and key play primitives to help players achieve game goals.

Storytelling for interactive entertainment, such as in computer games, differ significantly from stories written for linear media, such as films, plays, and books. However, the story structure itself can transcend different media and can be applied to both linear and interactive narrative. The difference is that one accommodates the audience's participation and choices to increase emotional engagement while the other does not.

Most interactive entertainment experiences follow a three-act structure with a beginning, middle and end. Act 1 is written to capture the audiences' attention with a compelling premise. The audience is often placed in an "ordinary world" of the protagonist to familiarize them with the lay of the land and to develop empathy for the main character, when suddenly an inciting incident turns the character's world upside down and propels the story forward. Act 2 advances the plot, providing the main character/player with escalating conflicts and a core dilemma from where they will have to take on escalating risks and challenges or choices that evolve the character. Act 3 provides a climax where the protagonist is able to confront the antagonist to resolve the dilemma initially confronted in Act 1. This series of events provides an exhilarating emotional catharsis as the story comes to either a resolution or evolution that concludes with a happy or tragic ending.

Interactive games also need to follow an emotional story arc to satisfy the audience's expectation. The difference from a linear story is that the audience has agency as the player and the experience arc is "specified by rules and not events" (Crawford, 2003). Instead of a linear plot, it becomes a dynamic "metaplot" where the player participates in the unfolding sequence of problems and obstacles that develop characters and forwards with story. Instead of linear plot points placed by the author, the influence of the audience gains meaning through interactions with the game to compel and not disruptive the story or game flow. In both interactive games and stories, the challenge lies with the author to create a journey that reaches the same emotional waypoints as a linear plot, but the audience is allowed to reach them in their own way (Stapleton & Hughes, 2003).

To create engaging, interactive story events that are designed intentionally to facilitate learning, educators identify instructional events, associated with the instructional approach or strategy selected during concept development, and work with game developers during pre-production to integrate the strategy with sound game design structure by exploiting similarities and resolving differences in story and instruction.

Hannifin, Hannifin, Land and Oliver (1997) define "grounded design" as "...the systematic implementation of processes and procedures that are rooted in established theory and research in human learning" (p. 102). A grounded approach uses research and theory to make design decisions and optimize learning. It neither subscribes to nor advocates any particular epistemology, but rather promotes alignment between theory and practice (Hirumi, 2002). Table 6.6 outlines events associated with alternative instructional strategies "grounded" in learning and instructional research and theory.

Table 6.6 Primary Events Associated with Grounded Instructional Strategies

Adaptive Instructional Design (Schwartz, Lin, Brophy & Bransford, 1992)	**Collaborative Problem-Solving** (Nelson, 1999)	**Learning by Doing** (Schank, Berman & Macpherson, 1999)
1. Look Ahead & Reflect Back 2. Present Initial Challenge 3. Generate Ideas 4. Present Multiple Perspectives 5. Research and Revise 6. Test Your Mettle 7. Go Public 8. Progressive Deepening 9. Reflection and Decision Assessment	1. Build Readiness 2. Form and Norm Groups 3. Determine Preliminary Problem 4. Define and Assign Roles 5. Engage in Problem-Solving 6. Finalize Solution 7. Synthesize and Reflect 8. Assess Products and Processes 9. Provide Closure	1. Define Goals 2. Set Mission 3. Present Cover Story 4. Establish Roles 5. Operate Scenarios 6. Provide Resources 7. Provide Feedback
5E Instructional Model (BSCS, 2006)	**Problem-Based Learning** (Barrows, 1985)	**Case-Based Reasoning** (Aamodt & Plaza, 1994)
1. Engage 2. Explore 3. Explain 4. Elaborate 5. Evaluate	1. Start New Class 2. Start a New Problem 3. Problem Follow-Up 4. Performance Presentation(s) 5. After Conclusion of Problem	1. Present New Case/Problem 2. Retrieve Similar Cases 3. Reuse Information 4. Revise Proposed Solution 5. Retain Useful Experiences
Experiential Learning (Pfeiffer & Jones, 1975)	**Simulation Model** (Joyce, Weil, & Showers, 1992)	**Constructivist Learning** (Jonassen, 1999)
1. Experience 2. Publish 3. Process 4. Internalize 5. Generalize 6. Apply	1. Orientation 2. Participant Training 3. Simulation Operations 4. Participant Debriefing 5. Appraise and redesign the simulation	1. Select Problem 2. Provide Related Cases 3. Provide Information 4. Provide Cognitive Tools 5. Provide Conversation Tools 6. Provide Social Support

During pre-production, an instructional strategy may be applied at the game or scene level to guide the design and sequencing of key events that occur throughout the game's story structure. For example, at the game level, the seven events associated with the learning by doing instructional theory (Schank, Berman & Macpherson, 1999), matched the archetypal action game framework that is being used to guide how the story is being written for an instructional game, currently titled *Trainien*[1]. Application of the seven event helps provide the bases for dramatic build up for the story all within context of the game.

1. **Define Goals:** The story and game structure is defined by establishing clear and compelling goals. For *Trainien*, the instructional goal was defined as, *"Given alternative training and educational situations, instructional designers will work effectively with game developers to systematically analyze, design, and test instructional games."* The game goal was specified as, *"Kick butt so the player can either fulfill his or her dream to work for the game company, run the education division, and hang out with cool creatures while playing with holodeck, or get home."* These goals were then embedded within a structured game premise presented to players as a mission.
2. **Set Mission:** In our case the protagonist (aka. player or learner) is abducted by alien game developers. His or her mission is to navigate through the alien world and either return home or remain, to work for the game developers, but she or he can only do so after assimilating the game development process, understanding the alternative roles and goals of game developers, and applying pedagogy during the instructional game development process. The mission is made less overwhelming by breaking it down into levels and scenarios. Each level of *Trainien* sets up smaller tasks to successfully achieve the final mission and ultimate learning goal by exploring each facet of the game development process with a wide variety of options and tools.
3. **Present Cover Story:** The emotional investment and interactivity that enhances game play often happens in the backstory. This event can allow us not only to embellish the story, but also establish appropriate choices and levels of difficulty within the set-up of the game to make the experience relevant. For example, the selection of rank, tools and resources during the presentation of the cover story provides players with the ability to make the game experience their own. In *Trainien*, the player is placed in their desired professional position (as an instructional designer) where they will need to learn from experience or "trial by fire." By placing the player in a popular fictional alien theme, real professional situations can be rendered with humor and drama to reduce the student's fear of failing and create a memorable experience. This encourages

---

[1] Trainien is being designed to augment a graduate level course on instructional game design. The basic idea is to provide a game that mimics the game development process. The game is not meant to replace the course; rather, to enhance the course experience with relevant examples of course topics and issues embedded within the game play while modeling effective game design.

players to be proactive in an unfamiliar work environment and feel comfortable with learning from their mistakes.
4. **Establish Roles:** Point of View is a critical element for dramatic writing and mission rehearsal. The ability to not only select our own roles, but establish relationships with others helps establish the collaborative nature of the game play and lesson objective. In *Trainien*, we are helping learners develop their own skills and knowledge relative to instructional game design, so the first person point of view is more appropriate than an objective third person point of view. The extreme characterization of the other characters is to gain the appreciation for the diversity and dramatics of working on a creative game development team. These roles reflect both the professional discipline and the archetypal characters (protagonist, antagonist, etc.). This not only contributes to a good story but also provides strategic options for achieving the learning objectives.
5. **Operate Scenario:** Entertainment is about meeting expectations and games are about making failure fun (to motivate learners to get up and try again) and this allowed us to take full advantage of what is core to the strategy and the game. In *Trainien*, we designed scenarios for each game level so that mistakes are the most attractive option. However in creative endeavors, there is never just one right answer. The operation of the game is to make a series of strategic choices, culminating in drastically different end results that illustrate inter-relationship between choices. For example, game level zero (Game Tester's Hell) enables learners to distinguish game genres. In the scenario, the player enters a back area where drones are brought in to test games of various genres. The player watches and is tasked with capturing games of certain genres while avoiding detection. Hybrid games and trick genres are aimed at throwing off the player. After collecting what she or he feels are the correct games, the player returns for an inquisition (assessment) to see if she or he is worthy to live. If the player does not succeed, she or he will die a horrible and creative death. If the player is successful, she or he will earn a medallion to activate the next level.
6. **Provide Resources:** Besides clear goals, cool tools can help make the game successful. This includes the virtual resources available that not only help with the incentive of the game, but become a persistent way to monitor progress, provide feedback and foster critical thinking. A key resource is access to intelligence or knowledge. In *Trainien*, an InterPlay Digital Assistant (iPDA) will enable players to gain and store clues, as well as communicate with mentors (e.g., the instructor or game characters, as well as real and fictional classmates) for coaching and advice. The more they use the cues or advice, the more we understand the player's familiarity of the subject. If they don't use the cues or advice and still perform poorly, that tells us about the player's attitude or behavior.
7. **Provide Feedback:** The procedural process of the interaction provides the constant tracking of choices and monitoring of performance. However, in games it is more about how you deliver the feedback than how you measure. The three recommended feedback types provide a variety of interactive mechanisms to

deliver it from *consequence* (making failure fun and immediate), *coaching* (a hero's mentor providing just in time advice), and *domain experts* (role-playing opportunities to explore the domain). *Trainien* utilizes all three options for feedback and is able to monitor the player's preference by the choices they make. This helps adapt the game to the learner's needs and interests.

The same learning by doing instructional strategy, as well as alternative grounded strategies, may also be applied at the scene level within the overall story structure. For example, the five events associated with the BSCS 5E model (BSCS, 2006; Bybee, 2002) are being integrated with the scenario operation of one game level within *Trainien*. In the scenario, the player needs to emulate the operations of the character creation workshop of the alien ship. The player is confronted with a mission to **engage** their interest or lose their life. To achieve the objective, players must **explore** and survey the infrastructure of game development. As the player interacts with obstacles, challenges and other characters, they need to **explain** what they know in action and in words to keep their cover and progress through the level (building drama). When appropriate, they will begin to **elaborate** on their own game creation recruiting the characters they have met or tools they used to be **evaluated** by the Supreme Commander where they will succeed or perish.

The application of grounded instructional strategies, however, does not fully utilize what we know about teaching and learning to optimize game-based learning. Similar to applying a grounded strategy within a game level (or instructional unit), it is believed that the selection and integration of grounded instructional events may help to further optimize game-based learning.

Research suggests that different external conditions, or events, should be used to promote different types of learning (c.f., Gagné, 1977). Table 6.7 lists specific instructional events, compiled by Smith and Ragan (2005), that have been found to facilitate the learning of verbal information, concepts, rules, problem solving, cognitive strategies, attitudes and psychomotor skills.

Table 6.7 Grounded Instructional Events that Facilitate Achievement of Various Learning Outcomes

Learning Outcome	Grounded Events
Verbal Information	Associational Techniques
Names, labels, facts or a collection of propositions.	• *Mnemonics Devices* (e.g., "FACE" for "Every Good Boy Does Fine").
	• *Metaphoric Devices* (e.g., "white cells attack infections like soldiers attack enemy").
	• *Instructor or learner generated images* (e.g., pictures, graphs, tables and maps).
	• *Rehearsal* (e.g., Drill & Practice).
	Organizational Techniques
	• *Clustering and chunking into categories* (e.g., periodic table).

Learning Outcome	Grounded Events
	• *Expository and narrative structures* (e.g., chronologies, cause and effect relationships, problem solutions, comparisons and contrasts).
	• *Graphic and advanced organizers* (e.g., concept tree linking new to prior knowledge).
	Elaboration Techniques
	• Write *meaningful sentences* (e.g., sentences using elements of periodic table).
	• *Devise rule* (e.g., describe why elements are organized in rows and columns).
Concepts A set of objects, symbols or events grouped together on the basis of shared characteristics which can be referenced by a particular name or symbol.	• *Inquiry Approach* (e.g., exploratory or discovery learning that typically begins with a presentation of examples and non-examples of a concept.
	• *Expository Approach* (begins with an explanation of a concept and its key attributes).
	• *Attribute Isolation* (points out the critical attributes of a concept).
	• *Concept Trees* (hierarchical, graphic representations of a specified concept that illustrate the concept relationship to subordinate and superordinate concepts).
	• *Analogies* (supplied by instructor or generated by learners)
	• *Mnemonics* (when verbal information is important to concept learning or for helping learners remember the key attributes of a concept
	• *Imagery* (a mental image of concrete concepts, such as pictures, graphs, tables and maps presented by the instruction or generated by learners).
Rules Relational rules or principals and procedural rules or procedures.	Procedural Rules (Procedures)
	• Learn to *determine if/when procedure is required.* Provide correct answer feedback with learner controlled explanatory feedback.
	• Learn to *list the steps in a procedure.*
	• Learn to *complete the steps in a procedure.*
	• Learn to *elaborate sequence*, starting with simple epitome of rule and elaborating to more complex versions of same rule.
	• Learn to *check appropriateness of completed procedure.*
	Relational Rules (Principles)
	• Ask learners to create their own *mnemonic device(s)* to support principle
	• Ask learners to create *images/diagrams* that illustrate relationships of concepts as presented in the principle
	• Practice *stating principle* (in own words).
	• Practice *recognizing situations* where principle is applicable.
	• Practice *applying principle* to predict, explain, or control for effects of one concept on another.

Learning Outcome	Grounded Events
	• Practice *determining if principle was applied correctly*.
Problem Solving Combine learned principles, procedures, verbal information and cognitive strategies in a unique way within a domain to solve original problems	• *Presentation of the Problem* (case studies, simulations, limiting the number of rules–principles and procedures–that must be used, presenting explicit representations of necessary rules as cues, providing solutions to parts of the problem, limiting the amount of extraneous information). • *Problem Space* (Review directions and identify relevant information about goal state; Delineate and analyze relationship between current and goal states; Discern patterns; Define what is known and unknown about the problem and determine what information must be acquired to solve the problem; Break down the problem into intermediate states or subgoals). • *Appropriate Principles* (guided questions–generative approach–or direct statements–supplantive approach–on how to select and apply appropriate principles and procedures to move from the given state, through intermediate states, to the goal state. • *Practice* (Present multiple representations of the problem; Recommend techniques for limiting alternative approaches to problem resolution; Provide clues about the general form of the solution; Recommend search strategies for acquiring relevant information; Outline generic approaches for problem resolution such as hypothesis testing and working backwards; Establish criteria for evaluating the appropriateness of alternative solutions).
Cognitive Strategies Internally organized skills whose function is to regulate and monitor the utilization of concepts and rules	• *Discovery and Guided Discovery* (involves more direct instruction than discovery, helping learners ascertain particular strategies through the application of questioning strategies. • *Observation* (observe a model demonstrating the use of the strategy by paired, cooperative learners, expert demonstration; and symbolic visual or textual representation by fictional character • *Guided Participation* (Instructor works with learners to determine characteristics of learning task, identify strategies to facilitate the task, and determine effective methods for employing the strategy • *Direct Instruction* (Identify utility of the strategy; Provide overview of steps and their relation to overall strategy; Demonstrate or model the strategy; Illustrate examples and non-examples of strategy use; Practice application of the strategy across gradually more difficult situations; Provide corrective feedback; Encourage and guide transfer of strategy to separate but appropriate context).
Attitudes Choice behaviors that make certain classes of action more or less probable	• *Demonstrate* desired behaviors representative of target attitude by a respected role. • *Practice* desired behavior associated with the desired attitude is another powerful tool in attitude formation and change (e.g., role playing and group discussions) • *Provide reinforcement* for the desired behavior (a stimulus that increases the probability of the preceding behavior reoccurring.

Learning Outcome	Grounded Events
	• *Communicate persuasive messages* from highly credible sources
	• *Create dissonance* (persuading learner to perform an important behavior that is counter–dissonant–to the person own attitude, attitude change may result.
Psychomotor Skills Coordinated muscular movements that may be difficult to distinguish from intellectual skills	• *Massed versus Spaced Practice* (massed practice engages learners in one or a few intensive periods of practice. Spaced practice exposes learners to short practice sessions distributed over time.
	• *Whole versus Parts Practice* (whole practice is advisable if the task is simple, not meaningful in parts, made up of simultaneous performed parts and has highly dependent parts, and if the learner is able to remember long sequences, has long attention spans and is highly skilled).
	• *Progressive parts practice* (if learners may have difficulties putting the parts together into a meaningful and well executed whole).
	• *Backwards chaining* (learners exposed to and practice the last step and work their way to the first step.

As game developers detail the story (related to questions), they should consider the terminal, instructional objectives for each unit or game level, and consider integrating one or more grounded events that have been found to facilitate the achievement of the particular type of objective. For example, the terminal objective for one level in *Trainien* is to distinguish key player and learner characteristics. To facilitate such concept learning, players will be presented with *imagery* (pictures that depict key attributes of people who are considered part of Generation X, Generation Y, the Net Generation, and the Millennial Generation) as the progress through the level and related story and game play.

As developers complete game design documents and generate a tangible prototype, it is critical for educators to verify that grounded instructional strategies and events are embedded within the artistic story and game play. A description of the instructional strategy may be too cumbersome to include in the design document verbatim; however it is fine to have references to the instructional strategy in the addendum. The key is to assimilate the strategies and events within the story and game play detailed in the game design document, so they can not be ignored as a working prototype is developed to verify game design.

Begin Formative Evaluations. Educators frequently implement initial drafts of instructional materials. In such instances, problems often occur and either the instructor is blamed for poor teaching or learners are blamed for their lack of attention or insufficient studying, when, in fact, the instructional materials were not well designed. To address this issue, Cronbach (1975) coined the term, "formative evaluations" to describe the collection and evaluation of data during development to improve instructional effectiveness and efficiency. Dick, Carey and Carey (2005) describe four types of formative evaluations, including expert reviews, one-to-one evaluations, small group evaluations and field tests. Expert reviews and one-to-one formative evaluations are encouraged before the tangible prototype is demonstrated to sponsors and other key stakeholders.

Expert reviews include formative evaluations by subject matter experts as well as media, learning and human factors specialists. Subject matter experts (SMEs) comment on the currency, accuracy and adequacy of the information provided within the instruction. Learning specialists review the objectives, instructional strategy and assessment methods. Media specialists are particularly important when developing audio, video, graphics and animations. Human factors experts examine the usability of the interface and navigation schemes.

The purposes of one-to-one evaluations are to identify and remove the most obvious errors in the instruction and to obtain initial reactions to the content from learners. During this stage, designers work directly with the person evaluating the materials. Direct interactions distinguish this phase of formative evaluation from others. Observation forms, interview questions, attitudes surveys and achievement tests are frequently used to gather data during this phase. Before achievement tests are used to evaluate student learning, they should also be formatively evaluated to help ensure validity and reliability.

Participants for the one-to-one evaluations typically include one learner who is above average in ability, one who is average, and one who is below average. The instruction is revised after each evaluation and designers may choose to conduct additional one-to-one evaluations if necessary. The primary criteria during one-to-one evaluations include: (a) clarity (is the message, or what is being presented, clear to individual learners?), (b) impact (what is the impact of the instruction on individual learner's attitudes and achievement of the objectives and goals?), and (c) feasibility (how feasible is the instruction given the available time, facility and material resources?)" (Dick, Carey, and Carey, 2005).

One-to-one formative evaluations should be integrated into the testing of the prototype. As the digital prototype is tested to ensure the game is fun and compelling, developers should also interact directly with the testers to evaluate clarity, impact and feasibility. Obvious errors should be removed and feedback from the testers should be incorporated, and the prototype revised after each evaluation.

The degree to which a selected instructional approach is successfully integrated with preliminary game design during concept development will affect the level of assimilation of instruction strategies and events within the design phase. The problem with many instructional game development processes is that pedagogy is not addressed until the pre-production phase and learning becomes superficial or secondary to the entertainment or vice versa. Constant reviews and documentation such as notes, suggested refinements, redline corrections, and punch-lists, along with expert reviews and one-to-one formative evaluations are tools for educators to insure proper application and integration of pedagogy within the game during pre-production, before the game goes into production.

6.4.3 Production Phase

After the prototype is approved, game developers enter the longest phase—production. During production, Alpha and Beta versions of the game are developed and tested before a final "Gold" version is delivered to the manufacturer for duplication, sales and marketing.

For the Alpha version, the game is playable from start to finish, but there may be few gaps and the art assets may not be final, but the engine and user interface are both complete. Each module is tested at least once and a bug database is created. Production of the Beta version focuses on fixing bugs and the integration of all assets. The objectives are to complete testing (including use on all supported platforms), bug fixing and performance tuning. Game elements, such as the code, content, path navigation, user interface, art, and audio, must be complete to pass the Beta testing. Once the game has passed Beta testing, it is considered Gold. Senior management has reviewed the product and the bug database and agrees that the product is ready for manufacturing. Master game discs have been thoroughly tested and the game is packaged for release into the marketplace.

Game developers are used to testing prototypes, during pre-production, and various versions of games during production. However, such tests tend to focus on ease of use, documenting and fixing programming bugs, and ensuring the game is (still) fun to play. Additional formative evaluations are posited as an integral part of production to eliminate remaining problems with, and enhance the pedagogical effectiveness of an instructional game.

Complete Formative Evaluations. To complete the formative evaluation process initiated during pre-production, additional evaluations are posited during production to assess the instructional clarity, feasibility and impact, and improve the pedagogical effectiveness, efficiency and appeal of instructional games. Specially, small group evaluations are recommended for the Alpha version of the game, and field tests are recommended for the Beta versions of the game.

The purposes of small group evaluation are to determine the effectiveness of changes made following the one-to-one evaluation, to identify any remaining learning problems, and to determine if learners can use the instruction with little to no interaction with designers. The basic procedure used for small group evaluations differ from the one-to-one evaluations in that after a giving a preliminary overview, designers administer the instruction in an environment that closely resembles its intended setting. Designers intervene only if equipment fails, or if for some reason, learners get stuck and cannot continue.

Small group evaluations are posited for the Alpha version of instructional games, when at least one game path is playable from beginning to end and the user interface is complete, but there may be a few gaps in the game play and the art assets may not be final. Each module of the Alpha version is typically tested at least once and a bug database and testing plan are created, including performance results. With the integration of small group formative evaluation methods, pretests may be used to measure learners' prerequisite skills and posttests may measure achievement of specified instructional objectives. Attitude questionnaires and follow-up interviews are also conducted, and the feasibility of the instruction is evaluated by estimating the time required by learners to complete the instruction.

This final stage of the formative evaluation is where designers attempt to apply the instructional materials in a learning environment that mimics its intended setting. The purpose of this stage is to decide if the alterations in the instruction made after the small group evaluation were effective and if the instruction can be used under targeted conditions. To answer these questions, all materials should be ready for use as intended. If an instructor is involved in executing the instruction, designers should not play this role.

Field trials are recommended for the Beta version of instructional games. The Beta version focuses on fixing bugs and the integration of all assets. With the integration of field trials using beta tests, the primary purposes at this stage of production are to complete testing, fix all bugs, fine tune performance, and ensure the game may be used in its intended setting. The game is basically complete, except for revisions made based on input gained from the trials. The procedures and instruments used during field trials should be very similar to that used for the small group evaluations. Instruments measure learners' attitudes and performance, and observations and interviews with learners and instructors are also valuable. The primary change is in the role of the developer. If the developer is not the instructor, the developer should only observe the process. The use of data gathered from field trials of the Beta version, along with final fixes to bugs in the code, results in the production of the Gold version of the game that is then released reproduction and packaging.

6.5 Conclusion

This chapter is written to help educators communicate and collaborate with game developers. It provides an overview of the game development process, and identifies common tasks completed, and deliverables generated during the process. It also discusses how, when and where educators can apply their knowledge of the subject matter, instructional situation and pedagogy during the process to optimize game-based learning, focusing on answering questions related to the three key components of interactive entertainment (i.e., story, game, and play) and their related elements.

As much of the discussion suggests, effective game design is based on the ability of developers to exploit similarities and reconcile differences in entertainment and education. In other words, successful game design depends, to a large extent, on the ability of team members to integrate and synthesize the results of instructional design and game design tasks in a creative, yet logical and systematic manner. Fabricatore (cited by Mitchell & Savill-Smith, 2004, p. 50) refers to this alternative design approach as 'edugaming,' where there is, "...no unnatural barrier separating learning from gaming" (Fabricatore, 2000, p. 14).

There are some who argue that instructional designers would "kill the fun" in instructional games, inferring the application of fundamental instructional design tasks has little to no value in game design. To some extent, we would agree. If left solely to instructional designers, with little to no knowledge of interactive entertainment and game design, resulting products may not utilize the potential of games to engage learners and make learning fun. However, we hope this chapter also illustrates the value of working with educators and instructional designers during the design and development of instructional games. We believe that sustained dialog and combined efforts of content experts and professionals in game design, simulation and modeling, software engineering, and instructional design, will lead to the development of motivating and pedagogically sound instructional games that utilize the potential of emerging technologies to optimize game based learning.

6.6 References

Aamodt, A. & Plaza, E. (1994). Case-Based Reasoning: Foundational Issues, Methodological Variations, and Systems Approaches. *Artificial Intelligence Communications*, *7*(1), 39-59. Retrieved March 15, 2005 from http://www.lai-cbr.org/theindex.html.

Barrows, H. S. (1985). *How to Design a Problem Based Curriculum for the Preclinical Years*. New York: Springer Publishing Co.

Bates, B. (2004). *Game Design* (2nd Ed.). Boston: MA: Thomson Course Technology, PTR.

Blake, J., & Goodman, J. (1999). Computer-based learning: Games as an instructional strategy. *The Association of Black Nursing Faculty Journal*, 10(2), 43-46.

Bybee, R. W. (2002). Scientific inquiry, student learning, and the science curriculum. In R. W. Bybee (Ed.). *Learning Science and the Science of Learning* (pp. 25-36). Arlington, VA: NSTA Press.

BSCS (2006). *Learning theory and the BSCS 5E instructional model*. Retrieved May 1, 2006 from http://www.bscs.org/library/Learning_Theory_and_the_BSCS_5E_Instructional_Model.pdf.

Crawford, C. (2003). *The Art of Interactive Design*. San Francisco, CA: No Starch Press, Inc.

Cronbach, I. J. (1975). Course improvement through evaluation. Reprinted in Payne, D. A., & McMorris, R. F. (Eds.), *Education and Psychological Measurement*. Morristown, NH: General Learning Press, 243-256.

Dempsey, J. V., Lucassen, B. A., Haynes, L. L., & Casey, M. S. (1996). *Instructional applications of computer games*. New York, NY: Annual Meeting of the American Educational Research Association. (ERIC Document Reprodcution Service No. ED 394 500).

Dick, W., Carey, L., & Carey, J. O. (2005). *The Systematic Design of Instruction* (6th edition), New York: Addison-Wesley Educational Publishers, Inc.

Driscoll, M. P. (1994). Psychology of learning for instruction. Needham Heights, MA: Paramount Publishing, Inc.

Egenfeldt-Nielsen, S. (2005). *Beyond Edutainment: Exploring the Educational Potential of Computer Games*. Retrieved October 15, 2005 from http://www.itu.dk/people/sen/egenfeldt.pdf.

Gagne, R.M. (1977). *The Conditions of Learning* (3rd ed.). New York: Holt, Rinehart, and Winston.

Hannafin, M.. J., Hannafin, K. M., Land, S. M., & Oliver, K. (1997). Grounded practice and the design of learning systems. *Educational Technology Research and Development, 45*(3), 101-117.

Hirumi, A. (2002). The design and sequencing of e-learning interactions: A grounded approach. *International Journal on E-Learning, 1*(1), 19-27.

Holland, W., Jenkins, H., Squire, K. (2002). *Video Game Theory*. In B. Perrron & M. Wolf (Eds). Routledge. Retrieved February 15, 2006 from http://www.educationarcade.org/gtt/

Jonassen, D. (1999). Designing constructivist learning environments. In C. M. Reigeluth (Ed.). *Instructional Design Theories and Models: A New Paradigm of Instructional Theory (pp. 215-239)*. Hillsdale, N.J.: Lawrence Erlbaum Associates.

Jonassen, D. H., Tessmer, M., and Hannum, W.H. (1999). *Task Analysis Methods for Instructional Design*. Mahwah, NJ: Lawrence Erlbaum Associates, Publishers.

Joyce, B., Weil, M., & Showers, B. (1992). *Models of Teaching* (4th ed.). Needham Heights, MA: Allyn and Bacon.

Kolb, D.A. (1985). *Experiential Learning: Experience as the Source of Learning and Development*. Upper Saddle River, NJ: Prentice-Hall, Inc.

Leake, D. B. (2000). *Case-Based Reasoning: Experiences, Lessons and Future Directions*. Cambridge, MA: MIT Press.

Leddo, J. (1996). An intelligent tutoring game to teach scientific reasoning. *Journal of Instruction Delivery Systems*, 10(4), 22-25.

Mitchell, A. & Savill-Smith, C. (2004). *The use of computer and video games for learning: A review of the literature*. London, England: The Learning and Skills Development Agency.

Nelson, L. (1999). Collaborative Problem-Solving. In C. M. Reigeluth (Ed.). *Instructional Design Theories and Models: A New Paradigm of Instructional Theory (pp. 241-267)*. Hillsdale, N.J.: Lawrence Erlbaum Associates.

Novak, J. (2005). *Game Development Essentials*. Clifton Park, NY: Thomson Delmar Learning.

Pfeiffer, J.W., & Jones, J.E. (1975) Introduction to the structured experiences section. In J.E. Jones & J.W. Pfeiffer (Eds.). *The 1975 annual handbook for group facilitators*. La Jolla, CA: University Associates.

Savery, J. R., Duffy, T. M. (1995). Problem-based learning: An instructional model and its constructivist framework. In B. Wilson (Ed.). *Constructivist Learning Environments: Case Studies in Instructional Design*. Englewood Cliffs, NJ: Educational Technology Publications.

Schank, R. C., Berman, T. R., & Macpherson, K. A. (1999). Learning by doing. In C. M. Reigeluth (Ed). *Instructional Design Theories and Models: A New Paradigm of Instructional Theory (pp. 161-179)*. Hillsdale, N.J.: Lawrence Erlbaum Associates.

Schwartz, Lin, Brophy, S., & Bransford, J. D. (1992). Toward the development of flexibly adaptive instructional designs. In C. M. Reigeluth (Ed). *Instructional Design Theories and Models: A New Paradigm of Instructional Theory (pp. 183-213)*. Hillsdale, N.J.: Lawrence Erlbaum Associates.

Sklar, M. (n.d.). *Mickey's 10 Commandments*. Retrieved May 01, 2006 from http://www.themedattraction.com/mickeys10commandments.htm.

Smith, P. L. & Ragan, T. J. (2005). *Instructional Design* (3rd ed.). Upper Saddle River, NJ: Prentice-Hall, Inc.

Stapleton, C. B. & Hughes, C. E. (2006). Believing is seeing: Cultivating radical media innovations. *Computer Graphics and Applications*. 26(1), 88-93.

Stapleton, C. B. & Hughes, C. E. (2003) Interactive imagination: Tapping the emotions through interactive story for compelling simulations. *IEEE Computer Graphics and Applications, 24*(5), 11-15.

Wikipedia (2005). *Game development*. Retrieved March 04, 2006 from http://en.wikipedia.org/wiki/Interactive_entertainment.

Wilson, B. (1995). *Constructivist learning environments: Case studies in instructional design.* Englewood Cliffs, NJ: Educational Technology Publications.

Chapter 7
Video Games and Teacher Development: Bridging the Gap in the Classroom

Elizabeth Simpson and Susan Stansberry

Abstract

The current focus on the research and development of video games and simulations for learning is not translating into widespread practice in classrooms. Researchers in this chapter provide tested models to anchor the use of video games and simulations to current structures of teaching activities that teachers are already comfortable with in order to bridge the gap between Digital Immigrant pedagogy and Digital Native students. Research meets practice as the authors describe the use of Lesson Study, a Game-to-Classroom Map, and Understanding by Design in successful efforts to support teachers implementing video games in the classroom.

7.1 Introduction

The gap between Digital Native students and their Digital Immigrant teachers has been well documented as the root of many problems in classrooms. (Prensky 2001, 2007; Gee 2005; Squire 2006). Research literature asserts that the use of video games can "stimulate the enjoyment; motivation and engagement of students, aiding recall and information retrieval, and can also encourage the development of various social and cognitive skills" (Mitchell & Savill-Smith 2004, p. 1). However, Digital Immigrant teachers encounter a variety of barriers in their efforts to bring video games into the classroom: lack of separation between different ways of using video games for learning, underdeveloped theory on facilitating learning through video games, weak theoretical knowledge of video games, and incomplete use of previous literature owing to the variation in terminology, place of publication, and researcher backgrounds. Preservice and inservice teachers need strategies for thinking about pedagogy in ways that can result in successfully bridging this gap in the classroom.

The majority of today's teachers are women (79%) with an average age of 46. In some states, 60% of the teachers are over 50 (National Center for Educational Statistics, 2003). Most of the experience these teachers have had with technology is limited to word processing, databases, presentation software, and perhaps some multimedia such as video and digital cameras. When these teachers bring technology into the classroom, they bring what they are familiar with and what they can control in a traditional teacher-directed lesson. Although this is a step in the right direction, it is not in itself engaging for the students.

> "Prensky (2001) describes the Games Generation as a group of workers who have solved daily mysteries (*Blues Clues, Sherlock Holmes*); built and run cities (*SimCity*), theme parks *(Roller Coaster Tycoon)*, and businesses (*Zillionaire, CEO, Risky Business, Start-up*); built civilizations from the ground up (*Civilization, Age of Empires*); piloted countless airplanes, helicopters, and tanks (*Microsoft's Flight Simulator, Apache, Abrams M-1*); fought close hand-to-hand combat (*Doom, Quake, Unreal tournament*); and conducted strategic warfare (*Warcraft, Command and Conquer*) – not once or twice, but over and over and over again, for countless hours, weeks and months, until they were really good at it." (p. 38)

Not only has this generation been raised with different tools for learning and play, research shows they also think differently than other generations. Moore (1997) referred to the Games Generations' ability to leap around in their thinking as "hypertext minds." I (Stansberry) saw this clearly when, in 2000, my 2-year-old son was frantically poking at the page of a hard back book while I was reading to him. He explained in frustration, that he wanted to "click on that picture to get more," but it wouldn't click. His hypertext mind was clearly at work at a very young age; it was unfathomable to him that he could not access non-linear material even within a linear media.

As a teacher of this generation of digital natives, it is imperative to understand the differences in them as learners. Table 7.1 shows a Prensky's (2001, p. 52) comparison of ways different generations learn.

Table 7.1 Learning Differences between Games Generation and Other Generations

Games Generation	Other Generations
Twitch speed	Conventional speed
Parallel processing	Linear processing
Graphics first	Text first
Random access	Step-by-step
Connected	Standalone
Active	Passive
Play	Work
Payoff	Patience
Fantasy	Reality
Technology-as-friend	Technology-as-foe

What the digital immigrant teacher may think of as fun and interesting and as a favor to the students (such as bringing in a video to class) is actually quite boring to those who crave interaction. The world of a digital immigrant is one in which he or she is free to actively connect with others in a random manner, while individualizing choices along the way.

Teachers are encouraged to embrace differentiated instruction, a method of meeting the needs of the individual learner, yet many teachers are unsure how to achieve differentiation while still maintaining control, thus differentiation is slow to be integrated into the system. Differentiation is inherent in the digital native's world. They make choices on a daily basis in every venue from food choices, information to be gathered, games to play, products to buy, and shows to watch, to cell phone ringtones, music, friends, and styles; the list is endless. If they want something they don't have, all they have to do is "Google it" and they are presented with a menu of choices of where to get it, how much it will cost, the history of the item and related subjects and sites. The environment is always individualized for them. This is the "have it your way" generation, a description emphasizing individuality, self-expressionism, and mass customization in today's pop culture (McCarthy, 2005). The "do it my way" philosophy of the traditional, old-school teacher-centered instruction does not work for this generation. A greater understanding of those learning principles will make it possible to incorporate them into the classroom environment. The hardware and software that make these games run is not as important as the cognitive skills players acquire while playing. How can we challenge and motivate our students in the same way games do, keeping students wanting more? What is left for educators to do is to extract the learning principles from games and marry them to the existing curriculum. In Table 7.2, let's revisit Prensky's (2001) description of learning differences and examine learning principles that teachers can use to differentiate instruction for their students.

Table 7.2 Merging Learning Differences and Learning Principles for Improving Classroom Environments

Digital Native Learning Differences (Prensky, 2001)	Learning Principles (Oblinger, 2004)	Incorporation into Classroom Environment
Twitch speed vs. conventional speed	Individualization	Learning is tailored to the needs of the individual. Some students work very quickly and are able to re-focus attention from one topic to the next instantaneously. This is a vast difference from keeping all students in one classroom on the same page of the same book on the same day.
Technology-as-friend vs. technology-as-foe	Feedback	Immediate and contextual feedback improves learning and reduces uncertainty. Technology is a friend, because it functions as a partner for most students, offer-

Digital Native Learning Differences (Prensky, 2001)	Learning Principles (Oblinger, 2004)	Incorporation into Classroom Environment
		ing feedback and help on a just-in-time basis.
Active vs. passive; Random access vs. step-by-step	Active learning	Engaged learners are able to concentrate on discovery and constructing new learning. With the ability to access learning from the point that quickly engages them, students are less likely to tune out directed steps that may be either too easy or too difficult for them to understand.
Graphics first vs. text first; Fantasy vs. reality	Motivation	In play, multiple senses are engaged. Visual elements draw the learners' attention and improve understanding of symbols. Fantasy motivates students, allowing them to think of themselves in a variety of roles and situations.
Connected vs. stand-alone	Social	Learning is a social and participatory process. Digital learners have clearly expressed their preferences for participation in social networking activities like MySpace, FaceBook, and collaboratively constructed "cheat" sites for games. Collaborative grouping in the classroom and online tools can facilitate digital natives' need for learning within a social environment.
Parallel processing vs. linear processing	Scaffolding	Games are built with multiple levels, so as the learner plays, they are gradually challenged with increasing difficulty of knowledge or skill. Additionally, previously learned information must be relied upon to solve problems. Without activating prior learning and drawing upon all content knowledge resources available, it is difficult to succeed at most games.
Play vs. work	Transfer	The ability to transfer learning from one context to another is a critical skill. When learners engage in what they consider to be play, they increase their enjoyment and increase their willingness to apply one learned concept to another situation.
Payoff vs. patience	Assessment	Learners need quality and timely feedback that allows them to learn through trial and error. Assessment should focus on the process of learning and the strategies individual students employ for problem solving rather than on rote memorization of facts without a context.

7.2 Use of Video Games in the Classroom

Beck and Wade (2004) conducted a large scale study of 2,500 business professionals to determine whether the experience of gaming and growing up surrounded by games, changes attitudes, expectations, and abilities related to how the video game generation performs in the business world (p. 21). The results indicate that gamers see the world very differently than do their parents, teachers or other non-gamers. The structure of the game molds the gamers' experiences, leading different expectations from learning environments and problem solving situations. In brief, Beck and Wade identified several ways in which digital natives are unique in their expectations. Prensky (2004) has likened the skills of the gamer to that of the highly effective workers of Stephen Covey's era. Gamers believe, for example:

7.2.1 There is always an answer

Video Games are basically fair. There is always a problem or problems that have a solution or solutions, which lead to an end result–the object of the game. There can be many different routes to reach the solution, all of which count equally if you achieve the goal. The answer is rarely obvious. A correct answer will give you information which will be useful in reaching the goal, thus you must persevere to find a correct answer. The answer is always relevant.

7.2.2 Nothing is impossible

In any game, you have the tools and the talent to be successful on your own or you may connect with someone who has the information you need in order to move forward (collaboration).

7.2.3 Trial and error

It is proven through game experience that this is the fastest, most efficient way to learn. If trial and error does not work, you know where to find the necessary resources, and you can access them at will. You are in control of your own success or failure. If you persevere, you will achieve your goal. If you do not win, restart and try again. You will not make the same mistakes twice. Failure is a learning experience – an opportunity to take another path – not an end result as it is so often in schools.

7.2.4 Competition and collaboration

Competition is inherent in game structure. Gamers are always competing. Competition is the motivating factor. Competition does not eclipse collaboration; in fact, collaboration is often an integral part of furthering your success. Competition and collaboration are symbiotic rather than mutually exclusive concepts.

7.2.5 Roles are clear

In games, roles are clearly defined. You are the droid or the Jedi, the good guy or the bad guy, an employee or the boss. You choose your role and understand its limitations. You understand the rules and the tools at your disposal, and you are willing to take the risks. In schools, the roles are not as clear.

7.2.6 Gamers are autonomous

Merriam-Webster Online Dictionary (http://www.m-w.com/) defines autonomous as the quality or state of being self-governing, especially the right of self-government. Gamers feel they have the right to choose their own path and are confident in exercising that right.

7.2.7 Gamers dominate their culture

There is little or no attention paid to non-gamers. Most teachers are non-gamers. Non-gamers don't have a clue regarding the video game environment, thus they can't be of any help. Unfortunately, most gamers realize that teachers are not gamers. Work by Kurt Squire substantiates that assumption, see below.

Many researchers in educational technology are investigating the application of video game concepts and technology for teaching people of all ages from primary school to adult job training, using such learning theories as problem-based learning (Gredler, 2003; Simpson & Clem, in press), cognitive apprenticeship (Paz Dennen, 2005; Stockhausen, Lynette & Ziitat, Craig (2002), and self-determination (Deci & Ryan 2007). The current research in cognitive science on how humans learn best is supported by the learning principles incorporated into good video games (Gee, 2003). Game players learn the value and benefit of taking on a new identity, discovering the cycle of expertise, understanding how to systematically order problems to aid in solving them, and that good learning involves performing a skill often prior to being competent with that skill. The learning principles found in video games can be directly related to the learning that should be taking place in our schools.

7.3 Teachers' Barriers to Video Games in the Classroom

Despite the popularity of high quality games and the support of open-minded educators the broad acceptance of game technology into the classroom faces many barriers, (McLester, 2005). The authors discovered several barriers that must be addressed when supporting the infusion of commercial off the shelf games into a teacher's methodology in the classroom: accountability, research-based tools and methodology, administrative support for innovation, professional collaboration, teacher preparedness, and scaffolding new methodologies with existing practice. These barriers are discussed below.

7.3.1 Accountability

The current political mandates for public schools require institutional accountability as measured against state standards on criterion based assessments. The assessments are used to measure the quality of student learning, often reported as basic, proficient, or distinguished on constructs which are usually knowledge based and closely tied to the curriculum. Often the assessments are developed with the curriculum publishers to insure the assessments reflect the material presented. Wyoming's PAWS test was developed in conjunction with Houghton Mifflin, a major content/curriculum provider. The data from these assessments are also used to identify and quantify the quality of the instruction provided by the institution. This means, frankly that if the teachers have not prepared the students to be able to respond to very specific knowledge based content driven assessments, the school and their jobs could be at jeopardy. Video games are viewed by teachers as being an "unknown". When using COTS to support standards based instruction, winning the game does not, without mediation and specific instructional goals and supports, equate to advanced content related standard specific knowledge (Simpson, in progress). Thus, the questions of how learning will be assessed, how that learning will be reflected on the district assessments, etc. must be addressed.

7.3.2 Research-based tools and methodologies

Currently schools are encouraged to utilize research based curricular tools and methodology in order to support student learning. Begehetto (2003) defines "scientifically based" as "Persuasive research that empirically examines important questions using appropriate methods that ensure reproducible and applicable findings." Research on the use of video games in the classroom is emerging from a variety of directions. Richard Van Eck suggests that researchers pull together to define a common thread or direction and utilize scientific methodology to support the use of videogames as tools in the classroom. The problem is that video games can be used in the classroom in several ways, each way needing research to validate the effects. Egenfeldt-Nielsen, (2004) notes several barriers that show up in the research:

- Lack of separation between different ways of using video games for learning (i.e. de Freitas, 2005);
- underdeveloped theory on facilitating learning through video games (i.e. Kirriemuir & McFarlane, 2003);
- weak theoretical knowledge of video games (i.e. Mitchell & Savill-Smith, 2004); and
- incomplete use of previous literature owing to the variation in terminology, place of publication, and researcher backgrounds(i.e. Squire, 2002).

For example, the video game can be used intact as a complete curricular tool. These are games specifically developed for educational purposes such as Math Blaster, Leap Frog, and others. Educational games rely on direct instruction as a teaching method and recall as the main assessment method. Direct instruction is a methodology that most teachers are very comfortable with thus, "games" that have learning contained in such packages meet the needs of many teachers as well as lend themselves to research. The impact of these games can easily be measured in quantitative ways. Another use of the video game is as an anchor to a lesson. An anchor provides a common experience by which the students can then attach new information. For example a high school English teacher suggested bringing in Vietnam, The Tet Offensive(t) by Oxygen Interactive, as an anchor to engage students in a literature discussion about the book, The Things They Carried, by Tim O'Brien. The teacher felt if the students could actually "be" a soldier in the war they would be more apt to be able to understand the potency of O'Brien's work. A social studies teacher might use the same game to have the students experience the complexity of the decisions made within the war and the effect of those decisions on the outcomes. Each use would require specific research methodology and produce varied outcomes based on the fidelity of instruction as well as other variables within the classroom, such as if the students are gamers already and how gamers approach the instructional objectives as opposed to non gamers. Games can also be used as a way to practice skills or content in situations that have been acquired in the classroom. For example, the authors are currently working with a high school advanced first aid teacher who is using Code Red, 911 as a performance assessment after explicit instruction on the proper procedures when faced with a medical emergency. Researchers have to be careful to choose the correct methodology for the question they are seeking to answer.

7.3.3 *Administrative Support for Innovation*

As addressed above, the pressures on educators to conform to measures of student success based on standardized outcome expectations has led administrators to support curricular decisions which are made on research-based, board approved curriculum. If the school is meeting the needs of the students, there is no reason for change. However, given the characteristic of the digital natives, often student's needs are not being met. In order for administrators to support innovation or change, a need must be demonstrated. The administrator then has to be able to support the need for change or innovation to the stakeholders. Data that indicate a need for change can be based either on student academic success or other indicators such as disengagement as measured by behavior referrals, drop out rates, teacher attrition, and student apathy. Currently, we know that 88% of dropouts have passing grades but are still electing to leave school (Thornburgh, 2006)). These are not your typical at risk students, these are instructional casualties that offer a strong data point for the need for change. The administration is the instructional leader in a school. Some instructional innovations that respond to the needs of disengaged students, require first order changes that are gradual and subtle such as Positive Behavior Intervention Supports (PBIS) or Response to Intervention (RtI), two initiatives which are scaffolded onto existing practice. Neither PBIS or RtI can be considered innovative. Both are based on current practice organized into "tiers of service" in which students are moved from group instruction to more individualized instruction based on their response to existing curricular or behavior interventions. Innovation most often requires second order changes that are drastic and dramatic (Marzano et al., 2005). Given the traditional habits, values and attitudes of many teachers, parents and administrators regarding video games, bringing COTS into the classroom could be viewed as a second order change. If the administration determines that the school is going to adopt innovation that requires a second order change such as incorporating video games into the curricular offerings, then they must prepare the stakeholders by investing the necessary time and resources.

7.3.4 Professional Collaboration

Teacher education is designed to insure curricular and content competence. Preservice classes are designed to produce teachers who can "run their own classroom", as if it were a ship sailing alone in an uninhabited ocean. Student teachers are marked as proficient if they are observed teaching a class independently and having good "behavior management" while the mentor is out of the room. Administrators who value change and innovation must problem solve and find ways for teachers to share expertise. Professional Learning Communities (PLC's), the Lesson Plan Study model, and professional development which connects teachers to an extended community using blogs, WebQuests, and wiki's offer educational professionals a framework to share ideas which could lead to innovation. Video games and the potential of video games as tools in the classroom are reliant on collaboration as a big part of the learning experience, both for the student and the teacher. Given that COTS offer fertile opportunities for cross curricular instruction, teachers must be willing to collaborate with their peers to insure that the "bridge" or instructional objectives designed for supporting the game play are being maximized. Teachers must be willing to problem solve and collaborate with students during those "on demand" moments when the game raises questions that must be answered right away in order for the connections to be made.

7.3.5 Teacher preparedness

The most important stakeholders in the process of bringing change and innovation to the classroom are the teachers. Yet, as mentioned above, many of our teachers are not gamers. In fact, early research shows that those entering the field of education may have very limited experiences with video games as well. In a preliminary study of a representative group of sophomores across colleges at the University of Wisconsin, Kurt Squire (2006) found that those students who were in the College of Education were much less likely to identify themselves as "gamers" as were those sophomores enrolled in other colleges. The education majors self-identified as having the least experience with video games overall. Therefore, not only are practicing teachers not widely experienced with adopting video games in the classroom, but preservice teachers are not bringing gaming experience to the profession. Colleges of education and the profession may be attracting students who are good at playing "school" in a traditional sense rather than those who are motivated to learn with, create, and engage others in these rich learning environments.

7.3.6 Need to scaffold new methodologies to existing practice

Time constraints, lack of training and lack of teacher motivation greatly affect the lack of success games and the associated learning principles have experienced in the classroom. Although many educators have little experience with games, they must be willing to delve into the learning principles that players gain from playing.

In *What Works in Schools*, Marzano comments that the longer a teacher teaches, the more competent they become in "dominant" rather than "cooperative behaviors" with students negatively affecting student behaviors (p. 93). Negative student behaviors cited by teachers most frequently are apathy, truancy and belligerence (National Center for Educational Statistics, 2005). David Sousa, author of the book, *How the Special Needs Brain Learns*, feels the way our schools are currently set up could, in fact, be disabling for some students who, given a different learning environment, might otherwise thrive (Sousa, 2001). Teachers who are not sure how to teach students who are not responsive to their traditional methods may be over-referring students for special services rather than reflectively examining their teaching methods. Even upon reflection, most teachers are not familiar with new pedagogies that will positively affect student performance and engagement, possibly due to a lack of interaction with new technologies in general. Unfortunately, this is true not only for our current in service teachers but also for pre-service teachers.

7.4 Strategies for Successfully Integrating Video Games in the Classroom-Bridging the Gap

When bringing commercial off the shelf (COTS) or edutainment games into the classroom, it is helpful for teachers to anchor the innovation to what they are already familiar with, such as identification of student learning standards and lesson planning. The authors have been involved in three different projects that used a type of anchoring technique: 1) a large federally-funded project that used Lesson Study with control and experimental groups prior to creating and introducing to the classrooms a massive multiplayer online role-playing game (MMORPG); 2) a state-funded project in Oklahoma that assisted teachers in integrating COTS into standards-based content using a Game-to-Classroom Map, and 3) a state-funded project in Wyoming that employed *Backwards Design* to help teachers anchor the use of COTS and edutainment games in the classroom to standards and outcomes. The teachers were given a lesson planning format (G.A.M.E.) which directed them to examine the standard, first, then look at the game as a tool for helping students reach the standards. The project allowed teachers the opportunity to select games they felt were well matched to the standards, spend time in the game environment, then go back to their standards and build a lesson plan.

Lesson study (Takahashi 2000; Yoshida 1999; Lewis 2004; Lewis, Perry, Hurd & O'Connell 2006; Lewis 2002) is a process that originated in Japan and, since 2000, has become widely used in the United States. In lesson study, a group of teachers work collaboratively to plan, teach, observe, and analyze actual classroom lessons and student data, engaging in a continuous cycle of discussion and revision. Lesson study offers a mechanism for teachers to systematically investigate and improve their teaching, but it also is helpful in introducing an innovation into the classroom. The Star Games project (K20 Center, 2005) led by the K20 Center at the University of Oklahoma is using lesson study professional development with control and experimental groups of teachers. Both groups have spent two years incorporating lesson study into their teaching. This experience is designed to assist all teachers in looking at teaching and learning from a common perspective. A MMORPG operating on ultra-mobile personal computers (UMPC) was then introduced into the experimental group classrooms. In this instance, lesson study served as an anchor to secure a common pedagogical understanding prior to introducing videogames into the classroom. Since lesson study is a process designed to shift *teaching as telling* to *teaching for understanding*, the teachers are better prepared to embrace innovations (MMORPG in this case), and employ them effectively in the classroom.

The study of the Star Games project use of lesson study and a MMORPG is currently underway and involves three traditional experimental groups and one development group:

- Group 1: A treatment group of nine schools randomly assigned to participate in the lesson study professional development and introduce the game/simulation on UMPCs after practicing Lesson Study.
- Group 2: A control group of seven schools randomly assigned to participate in the lesson study professional development.
- Group 3: A second comparison group that receives only the groundwork K20 Center school improvement professional development. Participation in this professional development program was a prerequisite for the treatment and control schools.
- Group 4: Two schools selected to participate in software testing with developers.

Initial observations indicate that participating teachers' initial engagement in lesson study has provided a foundation for introduction of an innovation into the classroom.

Teacher reticence to widely incorporate video gaming in classroom instruction may be directly related to the tradition of teaching as telling, with the teacher considered to be the expert in all areas of classroom knowledge. When asked to use a video game in instruction, the teacher may first think they will have to become an expert in the game prior to introducing it to the students. As described with the use of lesson study, anchoring the teacher to what is familiar, yet taking them on a path toward teaching for understanding can facilitate effective integration of video games. The Game-to-Classroom Map (Figure 7.1) was created for this very purpose: to take teachers on a familiar journey of basic instructional design, but also to guide them in seeing the standards evident in a particular game. The Game-to-Classroom Map walks teachers through the following stages:

- Good Connections to Curriculum (standards)
- Abilities/Understandings/Resources needed (essential questions)
- Make it Happen (activities/objectives)
- Extending the Experience (assessment of learning)

Teachers participating in a Math, Science, and Technology Teacher Institute were asked to explore selected COTS games (e.g. *ZooTycoon* and *SimCity*) and complete a Game-to-Classroom Map while they explored. Through this guided exploration, teachers could see that their role as content expert could be coupled with students' expertise with gaming in order to increase student engagement with and understanding of content.

At the summer Math, Science, and Technology Teacher Institute, a "Game Room" area was set up with ten computer stations where teachers could relax and experience the following COTS games: *Age of Empires, Bioscopia, Civilization 3, Dimenxian, Evolver, Genius, Quest Atlantis, Real Lives, Roller Coaster Tycoon 2, Sim City 4, Trade Empires,* and *Zoo Tycoon.* To begin with, in a group setting, the workshop leader walked step-by-step through one activity in *Zoo Tycoon* (setting up four different habitats and placing appropriate animals and their needs in the correct habitat). The teacher participants were asked to identify during the walk-through any standards-based content or process skills they recognized, which they easily did. As a result, many of the participating teachers were able to see that games could connect to content they were already teaching. This was an important recognition, as many of the participating teachers said they would not bring video games into the classroom because they needed to prepare students for standardized tests. By clearly seeing content and process skills in a video game, the likelihood of the teachers' adoption of video games as an instructional tool should increase.

Fifty participating teachers responded to a survey designed to assess attitudes toward videogames. Responses to the question of how often they play video games (on PC, PS2 or Xbox, mobile devices or WWW) are shown in Table 7.3. only four teachers reported playing every day. Other responses included "at least once a week," eight teachers; "at least once every two weeks," five teachers; "at least once a month," seven teachers; "less than once a month," ten teachers; and "I never play," 16 teachers.

Game-to-Classroom Map

Game: Zoo Tycoon
Lesson: Creating a Habitat
Brief Description: The player handles all aspects of running a zoo. In this lesson they will create 3 habitats and add animals.
Grade Levels: 6-8
Teacher: Stansberry

Good connections to curriculum (Oklahoma Priority Academic Success Skills)

Abilities/understandings/resources needed (Essential Questions)

Science Process Standards
1 Observe and measure, 2 Classify, 4 Experiment, 5 Interpret and communicate

Life Science Content Standards
Structure and function in living systems 6.3, 7.2
Populations and ecosystems 6.4
Reproduction and heredity 7.3
Behavior and regulations 7.4
Diversity and adaptations of organisms 8.3

Math Process Standards
1 Problem solving
2 Communication
3 Reasoning
4 Connections

Math Content Standards
Number sense 6.2, 7.2, 8.2

Basic computer skills (click on icons)
Low reading level
Basic knowledge of animals and environment
Have any of the students played this game before?
What experience with zoos do the students have?

Teacher resources:
Teachers' Guide
http://www.brainmeld.org/TeachingGuideLibrary/BrainMeld-ZooTycoon-Hillman.pdf
Official website http://zootycoon.com

Make it happen (Activities/Objectives)

Extending the experience (Assessment of Learning)

1. Set up a station/center. Small groups of students rotate through this station in groups of three:
   -Mouse Driver – this group member controls the mouse and clicks where the group decides to go
   -Information Expert – gathers necessary information on game play (rules, tips & tricks) and/or content information
   -Recorder – keep notes on what choices the group makes. Notes can include text, pictures, drawings, etc.
2. decide which animals belong in each of 3 small zoo exhibits
3. adopt animals – use pop-up tips about each animal to best match needs/wants with environment
4. Use construction icon to add foliage, rocks, terrain appropriate to the animals' habitat(s)
5. Click on animals to monitor health and happiness
6. PAUSE game and do discussion

Discussion Questions:
1. What is an animal habitat?
2. What must be in an animal's habitat in order for the animal to survive and be happy?
3. What do all animals need to live?
4. How are the animal habitats you created the same? How are they different?
5. Could the animals in your zoo live in another animal's habitat? Why or why not?
6. Can some of your animals in your zoo live together in the same exhibit? Why or why not?
7. Is there anything you put into the habitats that animals can live without?
8. There are no zookeepers in the wild; how do the animals survive without a zookeeper taking care of them like they do in a zoo?
9. Write a story about your favorite animal habitat in your zoo. Describe the habitat and tell what animal lives there. Draw a picture illustrating your story.
10. Create a diorama

Fig. 7.1 Game to Classroom Map for Zoo Tycoon Lesson

Table 7.3 Teacher Responses to Frequency of Video Game Play

	Everyday	At least once a week	At least once every two weeks	At least once a month	Less than once a month	Never
Number of Teachers	4	8	5	7	10	16

Based on these numbers, this group of teachers was considered to be largely non-gamers. A majority of this group (31) also reported never discussing games with their students, while 15 did discuss games with their students and expected to continue that practice.

Upon completion of the Math, Science, and Technology Teacher Institute, participants were asked why they would now consider using videogames in the classroom for educational purposes. Seventeen participating teachers noted that they would consider using videogames in order to enhance an existing lesson, 14 to capture students' attention and/or keep their interest, 10 to increase students' thinking skills, and nine to make learning fun. Other reasons mentioned included improving knowledge retention, addressing different learning styles, building technology skills, creating competition, and measuring accomplishments. The purpose of bringing video gaming into this teacher professional development was merely to introduce participants to the idea of seeing content standards in video games and considering how why they could be incorporated into classroom instruction. The data collected indicates a successful experience with this group of teachers.

A similar study was conducted using information gathered as a result of a Teacher Quality Enhancement grant offered by the State of Wyoming awarded to the Albany County School District, in Laramie, Wyoming in the summer of 2006. The grant supported an extended professional development opportunity for 38 Wyoming teachers titled Motivating Digital Learners with Video Simulation Games. The Motivating Digital Learners workshops were organized around a lesson planning model "G.A.M.E." that was designed to scaffold exiting teacher understanding of the lesson plan process to components which would allow for the consideration of the unique needs of the digital learner. The G.A.M.E. model utilizes the well known lesson planning methodology of Wiggins and McTighe's *Understanding by Design* as a platform for scaffolding new components to lessons which include responsiveness to digital learners as well as the inclusion of video games as tools to support content and skill acquisition. Participants were inducted into a continuing learning community through the use of blogs, listserves, and WebQuests which extended the opportunity for learning, reflection, and sharing on the part of the participants. The workshops were delivered over five days, four hours of instruction and practice each day with additional homework and blogging each night. The participants tended to be highly experienced teachers with expertise ranging from elementary thru high school covering all content areas including Physical Education, Math, Science, Social Studies, English, Family and Consumer

Science, Medical Technology, one librarian and two special educators. Only eight had four years or less of teaching experience; twenty-three had over ten years of experience and seven had over 20 years of experience. Only five participants self – identified as being gamers. Most had no previous experience with video games other than to give them to their children or grand children as gifts. Similar to Lesson Study and the Game-to-Classroom Map, the G.A.M.E. process of lesson planning is designed for teachers to incorporate video games into the classroom through a process that encourages scaffolding to existing teacher practice, such as lesson planning, standards based instruction and standards based assessments. The G.A.M.E. model is developed for the unique characteristics of the digital native. The model is based on Understanding by Design (UBD) by (Wiggins & McTighe 2001) and begins with the end in mind: What understandings should my students develop? How will they demonstrate what they have come to understand? How will I ensure that the students have met the standards for learning required on summative assessments? The G.A.M.E. model modifies and extends UBD to help teachers consider the needs of the digital native as they design lessons based on the standards, using video games as a curricular tool.

The first step in introducing a new methodology is to help teachers understand the need, especially as it relates to the unique learning expectations of the digital learner.

- **Design instruction that leverages the same motivators that game designers use**. Game designers know that goals, emotional impact, and the promise of both competition and teamwork draw kids into game play. When players choose to consider playing the game, they will seek out more information. For example: What are the rules of the game? Do I have the necessary skills to engage? Do I have the tools to play this game, e.g., correct console, enough RAM, etc.? If the player feels confident at the end of this stage of information gathering, they will likely start playing. If not, they will pass. If they choose to pass on this game, they will find another game to play that more closely meets their needs for challenge and control.
- Design instruction that openly acknowledges the right of the student to make a decision to engage or to choose a different game. Charter schools are set up with this concept in mind: e.g., a student who loves music might pick a music school, etc. Differentiated instruction helps teachers accomplish the same type of autonomy in the classroom given the constraints of the standards and curriculum. Adding video games to the classroom is another tool to help us differentiate and give the learner choices. In game play, once the player has decided to play, they go through a cycle of steps with each move.
- Motivation is a key factor in game design that is often overlooked by teachers. Game feedback is normally immediate and often dramatic, supplying the last link in the motivational chain. Players look for satisfaction in their play; that satisfaction may come in the form of strategic advancement, more freedom for the next move, glorious victory, or even crashing defeat. Remember, even de-

feat is seen as a learning experience; they player has learned what *does not work* and can immediately reinitiate play, using the expanded knowledge just gained.

Teachers can use these same steps in reaching and teaching the digital generation. Doing so requires an acceptance that these learners are different from earlier generations and have unique requirements. The authors examined the components of game play and came up with the following components that teachers need to understand and design opportunities for their lesson.

The teacher, as the instructional leader responds to the gamer by identifying the goal of the activity, outlining for the student the standards that support the lesson as well as the expected learning outcomes. The teacher also explains how the tools, such as the video game, will be used to meet the goals (See Figure 7.2).

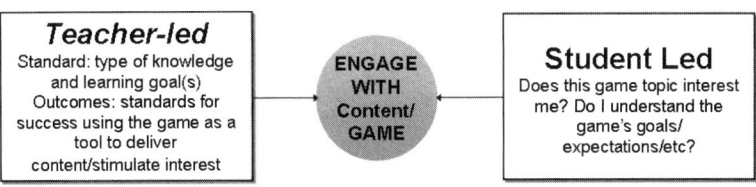

Fig. 7.2 Conditions for Engagement-Teacher Led Planning and Student Led Planning

Next, the teacher builds a lesson that will put the student through the same cycle of learning the student goes through when they are learning and mastering video games, the G.A.M.E. cycle.

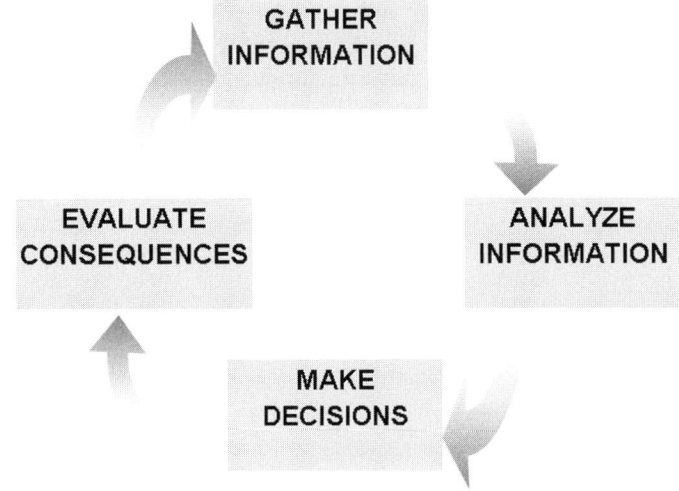

Fig. 7.3 G.A.M.E. Cycle

Table 7.4 shows the essential components to consider when lesson planning. The first column in Table 4 relates to video game play and the expectations of the gamer, the second column gives the expectations of the typical classroom, the last column explains how these components are wrapped into the G.A.M.E. lesson planning structure.

None of these approaches are novel; good teachers for years and years have been using similar processes to think critically about and plan carefully for student understanding. However, it seems that when faced with bringing an innovation into the classroom, teachers often abandon the basic instructional design processes. Therefore, an anchor like Lesson Study, Game-to-Classroom Map, or Understanding by Design, should be used to hold the innovation, the design of the lesson, and student learning steady.

In order to bridge the gap between Digital Natives and Digital Immigrants and overcome some of the barriers teachers face in teaching with video games, teachers and teacher educators need strategies for adjusting their pedagogical knowledge and skill. First, they need a better understanding of the unique learning needs and expectations digital natives have regarding learning environments. Recognition of their "immigrant status," is step one for a teacher ready to begin a journey of change. Video games in the classroom can force a change in the teachers' role in the teacher/student relationship from "dominant" to collaborative. Knowledge of learner expectations can help prepare teachers for that change. Secondly, teacher educators need to be familiar with the current research from medicine, business, sociology, and education showing that immersive video games are powerful learning tools in which the student can learn not only content, but 21st century skills needed to excel in an expanding world. Finally, teachers need to understand how bringing in video games will support the curricular expectations and the standards as well as enhancing the student learning experience. By anchoring the use of video games to current structures of teaching activities that teachers are currently comfortable with, teachers may begin bridging the gap between their pedagogy and their Digital Native students.

Video Games and Teacher Development: Bridging the Gap in the Classroom   181

Table 7.4 Lesson Plan Structure for Bringing Video Games into the Classroom

Meaning	Video Game	Traditional Classroom	G.A.M.E. Lesson Plan
GATHER The students gather information necessary to complete the task or goal. First information is gathered regarding the nature of the goal, second if the students feel the goal is relevant to their skills, they gather information necessary to achieve the goal.	At the beginning of play, the players gather information and analyze their options, what they know about the game and the expectations of the game. What is the goal of the game? What genre is the game? Is it similar to other games I have played? Will it challenge my skills? Based on what the player knows, she or he decides to engage, or disengage. Once inside game the player continues to gather information. What level am I on? What are the expectations of the level? What tools to I have at my disposal? What powers do I have? What are my obstacles? How do I move about? What are the objectives of the level? Etc. When designing a lesson plan for digital natives, it is critical that you have designed the lesson so the students can gather information regarding the standards you are going to be attempting to meet, the goal to be achieved, the tools available to them and how those tools can be used, the relevance of the goal to the student and how the student might utilize their unique skills to help the class achieve the goal.	The teacher knows the goals and objectives of the lesson but rarely shares that information with the student. Student choices are limited to rather to engage or not engage based on limited information. Attempts to gather more information "Why do we have to do this….what does this have to do with my life…etc." are often met with the teachers reprimands.	Introduces the game. The teacher determines the use of the videogame-anchor, support, content, etc. What information the videogame will support toward the goal. Guiding questions can be linked to the information gathered from the game to meet the objectives. Shares the goals and objectives with the students.

	Meaning	Video Game	Traditional Classroom	G.A.M.E. Lesson Plan
Analyze	Students analyze information against the goal to determine the usefulness of the information.	After gathering information players then begin to analyze the information they have gathered, making determinations on what is primary information and what is secondary to the objective	Analysis of the lesson is predetermined and measured against teacher determined right answers.	The lesson is designed with self-determination of the learner in mind. Such as- is there a process for students to reflect on the decisions they made and evaluate the consequences in the activities outside the simulation? Can students choose a different route or remediate their skills outside the simulation?
Make Decisions	Students make decisions regarding their analysis of the information and against the goal.	Once information has been analyzed, the information is then used to make decisions based on choices available. There are always choices in a video game, and the player must be ready to made immediate and well informed decisions or else they risk loosing the level or even ….dying!	Teachers make most of the decisions in the lesson. Students decide whether to engage or disengage.	Opportunities for student led decision making is built into the lessons. Decisions are either student generated and student led either alone or in collaboration with peers and experts or teacher mediated to insure consideration of agreed upon learning outcomes. Unanticipated decisions are welcomed and consequences considered.
Evaluate	External and internal evaluations/assessments give feedback to the student regarding their success in attaining the goal.	The decisions determined as necessary to reach the goal. Loosing progress toward the goal by loosing a level or dying is immediate and relevant feedback. The player then presses restart armed with information they learned by making an incorrect decision or not gathering enough initial information and begins the cycle again until the goal is accomplished	Evaluation is disconnected from activities and often seen as irrelevant.	Evaluation is designed that is purposeful, relevant and immediate. Since the goal of the lesson is to meet a standard the evaluation is directly tied to the standard being measured.

7.5 References

Annetta, L. A., Murray, M. R., Laird, S. G., Bohr, S. C., & Park, J. C. (2006). Serious games: Incorporating video games in the classroom. *Educause Quarterly*, *3*, 16-22. Retrieved July 7, 2007, from https://www.educause.edu/apps/eq/eqm06/eqm0633.asp

Beck, J. C. & Wade, M. (2004). *Got game: How the gamer generation is reshaping business forever*. Harvard Business School Press.

Carr, D. (2003). Game On: The culture and history of videogames. *Visual Communication*, *2*(2), 163-167.

Egenfeldt-Nielsen, S. (2003). In *Review of the research on educational usage of games*. Retrieved May 10, 2007, from http://www.it-c.dk/people/sen/papers/Reviewing%20the%20literature%20on%20simulations%20and%20games%20for%20learning_v0.5.doc

Engenfeldt-Nielsen, S. (2004). Practical barriers in using educational computer games. *On the Horizon*, *12*(1), 18-21. Retrieved Apr. 21, 2007, from http://www.emeraldinsight.com/info/about_emerald/emeraldnow/archive/emerald_now_pdf_archive/2740120104.pdf

Federation of American Scientists. (2006). *National summit on educational games fact sheet* Retrieved n.d., from http://www.fas.org/gamesummit/Resources/Fact%20Sheet.pdf

Gee, J. P. (2003). From Video Games, Learning about Learning. *The Chronicle of Higher Education, June 20, 2003*

Gee, J. P. (2005). In *Why are video games good for learning?*. Retrieved Feb. 12, 2007, from http://www.academiccolab.org/resources/documents/MacArthur.pdf

Gredler, M. (2003). Games and simulations and their relationships to learning. In D. Jonasson (Ed.), *Handbook of Research in Educational Communications and Technology, 2nd ed.* (p. 571-581).

K20 Center Star Schools Project, (2005). Retrieved Nov. 19, 2007, from http://www.k20center.org/okacts/phase3/star-schools/

Lewis, C. (2002). What are the Essential Elements of Lesson Study?. *The California Science Project Connection*, *2*(6), 1-2.

Lewis, C., Perry, R., & Hurd, J. (2004). A Deeper Look at Lesson Study. *Educational Leadership*, *February*, 18-22.

Lewis, C., Perry, R., Hurd, J., & O'Connell, P. (2006). Lesson Study Comes of Age in North America. *Phi Delta Kappan*, p. 273-281.

Marzano, R. J. (2003). *What works in schools: Translating research into action*. Alexandria, VA: Association for Supervision and Curriculum Development.

McCarthy, M. (2005, May 23). Burger king tries old slogan again. *USA Today*. Retrieved Dec. 5, 2007, from http://www.usatoday.com/money/advertising/adtrack/2005-05-23-burgerking_x.htm

McFarlane, A., Sparrowhawk, A., & Heald, Y. (2002). In *Report on the educational use of games*. Retrieved n.d., from http://www.teem.org.uk/publications/teem_gamesined_full.pdf

McLester, S. (2005). Game Plan. *Technology and Learning*, *26*(3),

Mitchell, A. & Savill-Smith, C. Learning and Skills Development Agency. (2004). *The use of computer and video games for learning: A review of the literature* London: Learning and Skills Development Agency. Retrieved Apr. 21, 2007, from http://http://www.lsneducation.org.uk/pubs/pages/041529.aspx

Moore, P. (1997). In *Inferential focus briefing*. Retrieved Sept. 30, 1997

Oblinger, D. G. (2004). The next generation of educational engagement. *Journal of Interactive Media in Education*, *8*, Retrieved n.d., from http://www.muzzylane.com/downloads/oblinger-2004-8.pdf

Pas Dennen, V. (2003). Cognitive apprenticeship in educatonal practice: research on scaffolding, modeling, mentoring, and coaching as instructional strategies. In D. Jonasson (Ed.), *Handbook of Research in Educational Communications and Technology, 2nd ed.* (p. 813-828).

Prensky, M. (2001). Digital natives, digital immigrants. *On the Horizon*, *9*(5), 1-6.

Prensky, M. (2001). In *Digital Game-based Learning*. (chap. Why education and training have NOT changed). Retrieved Sept. 17, 2007, from http://http://partners.becta.org.uk/page_documents/research/emerging_technologies07_chapter4.pdf

Prensky, M. (2007). How to teach with technology: Keeping both teachers and students comfortable in an era of exponential change. *Emerging Technologies for Learning*, *2*, Retrieved Sept. 17, 2001, from http://partners.becta.org.uk/page_documents/research/emerging_technologies07_chapter4.pdf

Prensky, M. (n.d.). In *Games Parents Teachers*. (chap. A parent-teacher toolkit). Retrieved Sept. 17, 2007, from http://www.gamesparentsteachers.com/

Prensky, M. (n.d.). In *Social Impact Games*. (chap. Entertaining games with non-entertainment goals (a.k.a. serious games)). Retrieved n.d., from http://www.socialimpactgames.com/

Shaffer, D. W., Squire, K. R., Halverson, R., & Gee, J. P. (2004). *Video games and the future of learning*.

Sousa, D. (2006). *How the special needs brain learns*. Thousand Oaks, CA: Corwin Press.

Squire, K. (2006). From content to context: Videogames as designed experience. *Educational Researcher*, *35*(8), 19-29.

Squire, K., Giovanetto, L., & Chmiel, M. (2006). Media literacy among undergraduates. *Association for Education, Communication and Technology*. Conference presentation?

Stockhausen, L. & Ziitat, C. (2002). New learning: Re-apprenticing the learner. *Education Media International*, *39:3*(4), 332-337.

Takahashi, A. (2000). Current trends and issues in lesson study in Japan and the United States. *Journal of Japan Society of Mathematical Education*, *82*(12), 15-21.

Thornburgh, N. (2006, April 17). Dropout nation. *Time*, *167*, 34.

Wiggins, G. & McTighe, J. (2001). *Understanding by design*. Prentice Hall.

Yoshida, M. (1999). Lesson study: A case study of a Japanese approach to improving instruction through school-based teacher development (Doctoral dissertation, University of Chicago, n.d.). *Dissertation Abstracts International*,

Chapter 8
Confronting the Dark Side of Video Games

Christian Sebastian Loh

Abstract

New video games have increasingly allowed players to choose their own destiny (or path of character development) by presenting them with choice of actions from both light and dark sides. Some studies suggested that players who have been exposed to the "dark side" of video games may act out unacceptable social behaviors based on their learning experiences. This issue of the effect of unintentional learning in video games deserves much scrutiny and honest discussion among researcher. Unfortunately, the issue has so far, either been bashed or glossed over by experts on both side of the effect-of-video-games chasm.

As educators, we have the responsibility to carefully evaluate and consider the effects of video games in learning because it is our duty to educate and shape the young minds. Rule #1 remains, "Do no harm!"

8.1 Introduction

The new millennium has brought about many exciting changes, amongst them a complete turnabout with regard to the initial perception of video games as an entertainment into a medium for learning. Although parents of children who spent many hours playing video games may still insist that playing video games is a 'time wasting' activity, a growing number of scholars, educators and scientists are coming to grips with the medium as a viable tool for learning (e.g. Feller, 2006; Gee, 2005; Gibson, Aldrich, & Prensky, 2007). In 2006 alone, the total revenues spent on gaming software and newly released game consoles (namely Xbox 360, PlayStation3 and Nintendo Wii) reached an unprecedented $12.5 billion (Ortutay, 2007). The Next Generation game consoles were so overwhelmingly in demand that the new devices were sold out on the very day of the product launch (Wallace, 2006)! It is widely believed that the multi-billion dollar electronic gaming industry will fuel a worldwide craving for video games for many years to come.

To meet the heightened demand for even more video games, game publishers are in need of a bigger workforce, including level designers, story/script writers, graphic artists, and game programmers in the coming years. More than 100

Colleges in the United States (U.S.) have stepped up to the plate by offering video game-related degrees to meet the growing market demand (Associated Press, 2005). Some major game publishers (such as Electronic Arts) have even awarded endowment money and internships to universities to help support their video game design courses (CNN Money, 2005). Many academics became interested in unique features of video games and how the new affordances may directly, or indirectly, affect learning. This is a marked departure from research efforts in the past that focused primarily on the negative effects of media on society.

8.1.1 Unique Features of Video Games

Despite the initial wave of approval, much research will still be required to examine the unique features of video games for learning. A second booster shot from the video game-based learning (VGBL) movement came from the Federation of American Scientists (FAS). In 2006, the FAS issued a joint call with the Entertainment Software Association (ESA) for federal funding to be made available (Cheng, 2006), for research towards harnessing the power of video games for learning (FAS, 2006b). The report by FAS listed several benefits for using video games in learning, including: "clear goals, lessons that can be practiced repeatedly until mastered, monitoring learner progress and adjusting instruction to learner level of mastery, closing the gap between what is learned and its use, motivation that encourages time on task, personalization of learning, and infinite patience" (FAS, 2006a).

There may be other media-specific features of video games that could enable a new way of interactive learning currently not possible. Some of these unique features include continual audio-visual stimuli, and 3-Dimensional (3D) virtual environments. Already, successful navigation of 3-Dimensional environments has been associated with the improvement of spatial cognition skills in gamers (Osberg, 1997; Sims & Mayer, 2002). Video games are an exciting and promising technology because they seem to 'have it all,' and at the same time, offer broad appeals to the 'gamer generation' (Beck & Wade, 2004), 'digital natives' (Prensky, 2001a), and even scientists, researchers and academies. Eventually, it is everyone's hope that school children will directly benefit from the VGBL. Will parents find their kids spending too much time playing video games? Prensky (2006) already has an answer: "Don't bother me Mom – I am learning!"

8.1.2 The Holodeck[1] Experience

The most fantastic feature about video games is its affordance[2], or visual cue to its function and use for players to step inside the *'magic circle'* (Salen & Zimmerman, 2004, p. 95) to interact with virtual people and objects within an imaginary *gamescape* (King & Krzywinska, 2003). Murray (1998) compared this feature to Star Trek's *Holodeck*, which she described as the *'ultimate fantasy machine'* in human imagination. An even more intriguing phenomenon happened within this video game *Holodeck*. Although players knew that virtual characters within a video game consisted of only colored digital pixels and shaded polygons, they often found themselves treating the virtual bodies as real people. One writer described, "You may know at one level of your mind [that these people] are colors on a screen, but your mind at another level sees them as people" (Mills, 2006). The meshing of reality with the virtual is looming as hardware technology and 3D modeling software become more advanced. Soldiers who engage in virtual battle training often cannot distinguish what's real and what's not (Karr, Reece, & Franceschini, 1997).

With current technology, it is already possible to create virtual humans with anatomically-correct bodies with freckles, feelings[3], and flowing hair. In the not so distant future[4], advances in auto-stereoscopic[5] projection displays (Weinand, 2005) and photorealistic computer-generated models with a full-range of emotions (Grizard & Lisetti, 2006; Paleari & Lisetti, 2006) will complete the illusion of immersive play (Brown & Cairns, 2004) — bringing the concept of *Holodeck* closer to reality.

---

[1] A virtually simulated "room" found on starships and star-bases in the *Star Trek* universe. Users, who stepped into the Holodeck, entered an immersive environment where they could interact with objects and people in it just like in the real world.

[2] Norman's affordance: http://www.interaction-design.org/encyclopedia/affordances.html

[3] These feelings are expressed as emotes (or actions that mimic human expressions of emotion). A virtual character's range of "emotes" is pre-determined by the programmer. For example, if the programmer did not create a "crying" emote, the game character will not be able to cry.

[4] Can't wait for the future? Check out http://www.vrealities.com/vrealviewer3d.html

[5] Auto-stereoscopic: Display of 3D (stereoscopic) images without the need of extra devices like shuttle glass or Head Mounted Display.

8.2 Video Game Playing

8.2.1 Practice Makes Perfect?

The common adage, "practice makes perfect," suggests that a person who devotes time to repeatedly performing a certain task will eventually achieve expertise. It should be of no surprise to anyone that a person who has chosen to dedicate many hours to diligently practice a certain cognitive-motor skill will eventually reach some sort of competency in the said skill. (Unfortunately, this belief might have also contributed, in one way or another, to the prevalent 'drill-and-kill' practice found in many classrooms.) When surveyed, professionals with a *culture of practice* (e.g. musicians, athletes, chess players, pilots) (MacMahon, Helsen, Starkes, & Weston, 2007) often attribute their success, or expertise, to copious *practice*. For instance, the amount of time spent on serious play for chess grandmasters was found to be nearly 500% more than that reported for intermediate-level players (Charness, Tuffiash, Krampe, Reingold, & Vasyukova, 2005). Compared with amateur pianists who spent an average of 1.88 hours per week in solo practice, award-winning pianists spent an average of 26.71 hours weekly (Ericsson, Krampe, & Tesch-Romer, 1993) — a difference of more than 1,420%!

Recent neurophysiology studies into video game players also appeared to support the notion of *practice makes perfect*. Findings showed (a) human brains secrete a substantial amount of *dopamine*[6] during video game playing (Koepp et al., 1998), (b) some video games can stimulate as much dopamine release as that induced by amphetamines (Kapur & Seeman, 2001, p. 364), and (c) *practice* stimulates the human brain to activate certain neuro-cognitive pathways that help enhance skill acquisition (Smith, McEvoy, & Gevins, 1999). Hence, Rosser Jr. and colleagues (2004) suggested that avid gamers who spend many hours each week playing video games will eventually master the video games (if not the game controller) through the establishment of new learning pathways in their brains.

8.2.2 Deliberate Practice

Decades of research study related to 'expertise and expert performance' likewise support the notion of "practice makes perfect" (Ericsson, Charness, Feltovich, & Hoffman, 2006). Ericsson, *et al.* asserted that 'expertise' was the result of *deliberate practice* and that "the highest levels of performance and achievement appear to

---

[6] Dopamine is a neurotransmitter that produces feelings of enjoyment ("reward") in humans. Thus it could reinforce or motivate a person proactively to perform certain activities.

require at least around 10 years of in-tense prior preparation" (1993, p. 366). The 10-years rule [7] was first observed in international chess masters (or grandmasters) and has been verified in several domains, including: music, music composition, mathematics, tennis, swimming, and long-distance running. Ericsson commented that surgeons are the only kind of medical doctors who get to improve their skills through deliberate practice: they set goals and obtain immediate and meaningful feedback with each successive surgery. Medical doctors, on the other hand, tend not to be able to receive feedback immediately (to confirm their diagnosis) and risk losing touch with their skills over time. This may help explain why laparoscopic surgeons who received additional motor-skill practices by playing video games were actually found to complete surgical procedures faster and with greater accuracy than their colleagues who received lesser or no additional training (Marriott, 2005). The investigators affirmed that video games used in this study were chosen specifically for their similarities to surgical procedures, which include: fine motor skills, reaction time, eye-hand coordination, non-dominant hand dexterity, two-handed choreography, targeting, and 2-D depth perception compensation (Rosser, Jr. et al., 2004). The implication is that: with persistence, sufficient practice, and appropriate feedback, almost anyone can become an expert of some sort: like Jonathan Wendel.

8.2.2.1 Fatal1ty

Born and raised in Kansas City, Missouri, Jonathan Wendel (better known as 'Fatal1ty') has risen through the ranks to become the top U. S. professional video game player. In an interview for CBS Broadcasting Network's '*60 minutes,*' he compared competitive video gaming to '*playing chess on caffeine*' (Court, 2006). He is a cyber-athlete, or e-sportsman, and would compete in cyberathlete (video game) tournaments just like other professional sportsmen (e.g., Tiger Wood and Michael Jordan). In just seven years (since 1999) of becoming a professional e-sportsman, he has won the World Championship in five different video games, and more than $1 million in tournament prize money. He has become the unofficial spokesman (or ambassador) for the e-sport industry and owns the license to a line of computer products and apparels with the brand: 'Fatal1ty.' His mission (or motivation) is to "prove that PC gaming is legit," both to the world and to his mother, who had objected strongly to his playing video games during his early years.

When he was traveling on tournaments, he spends whatever remaining waking hours practicing in the hotel room with a traveling 'sparring' partner. Even when at home, he tries to play for 8-12 hours a day because he treats video game playing

---

[7] The "10-year rule" was based on an earlier work (Simon & Chase, 1973), which documented that chess grandmasters required about a decade of intense preparation to reach that level of expertise.

as his full time job. He told the *Business Week* reporters (Hamm & Carney, 2005), "I set my goal. I wanted to be the No. 1 [sic] player in the world." He also stays focused on one game at a time so as not to be distracted. Once, he stayed on the game Quake III for 18 months! When asked about his secret to success, Wendel said, "Practice is the key to being the best at anything, but not everyone has the drive and determination… to become the best" (Cyberathlete Professional League, 2003).

8.2.3 Immediate Feedback

Interestingly, Fatal1ty's 'rising star' journey to become the video game grandmaster sounded conspicuously like *deliberate practice* (Dubner & Levitt, 2006), for he met all the conditions necessary for optimum learning and performance improvement (Ericsson et al., 1993):

1. *Be motivated* — Wendel is passionate about playing video games;
2. *Set clear goals* — he is determined to win tournaments;
3. *Adopt a training strategy* — he focused on techniques that would help him win: finding 'sparring' partners, practicing daily, staying fit, etc.;
4. *Obtain immediate feedback* — because of the way video games are designed, the gaming environment already provided him with the immediate and meaningful feedback he needed to improve himself; and
5. *Repeatedly perform the same or a similar task* — he engaged in video game practice for 8-12 hours daily.

Readers who are educators or instructional technologists may immediately recognize that the conditions listed above are reminiscent of several well-known educational learning theories, e.g. Robert Gagné's Nine Events of Instruction (Gagné, 1985), Robert Mager's Criterion Referenced Instruction (Mager, 1997), and John Carroll's Mastery Learning (Carroll, 1989). All of these theories have one thing in common: the importance of immediate feedback for effective learning. Educators have known for a long time that delays in feedback will hamper performance (Skinner, 1968). Ericsson echoes a similar belief, "In the absence of adequate feedback, efficient learning is impossible and improvement only minimal even for highly motivated subjects. Hence mere repetition of an activity will not automatically lead to improvement in, especially, accuracy of performance" (Ericsson et al., 1993, p. 367).

However, educators had to struggle with *immediate feedback* for many years, because it was humanly impossible to instantly provide feedback to 20-30 students in the class, or complete the grading of students' homework the moment they were handed in! The advent of computer technology, and computer assisted instruction (CAI) in the 70s, helped educators to overcome many of the hurdles of providing immediate feedback. Computing technology made it possible for learners to receive instant feedback about their performance. However, educators and instructional designers soon discovered the next hurdle: How many test questions can learners attempt before they become tired, bored, or irritable?

8.2.4 The Source of the Feedback

Video games do not have these limitations because they are designed to be fun (Koster, 2005), interactive (Crawford, 2003, 2004), and engaging (Michael & Chen, 2006; Poole, 2004) continuous play. In theory, a video game makes a great 'teacher' because once a player starts playing; the video game is immediately available to provide the player with a continuous stream of feedback! Feedback from scoring and audio-visual stimuli also allows learners to change or modify their learning strategies "before the ineffective ones become entrenched" (Javid, 2004). Yet, consider this: Who (or what) is the actual source of the feedback?

Because a video game is a software program, it is written by one or more programmers. Based on this argument, all stimuli and corresponding feedback must first be thought out and then coded into the game engine. Thus, it is up to the game makers (in short, everybody who has a say in what goes in to the game: programmers, publishers, level designers, artists, script-writers) to plan out and approve the video game's contents. In this case, the video game's contents includes a virtual world and its physics engine, player-characters modeling, landscapes, props, dialog, music, lighting, and missions, as well as the audio-visual stimuli and corresponding feedback. This brings the first dilemma in VGBL: **Who controls the video game's contents?**

DILEMMA #1: Who Controls the Video Game's Contents?

There is a saying that "those who have the gold set the rules." It is often the game publishing companies and sponsors who get to approve or decide the contents that go into video games. A majority of the people who design and develop video games do so as employees. They are likely to do what is being 'asked for' by the ones who control the purse-strings, because they are not likely to risk their livelihoods to do otherwise.

Example: *Under Siege* is a Middle-Eastern themed video game that targets Muslim children (aged 13 and above) by depicting them as the 'good guys' against Israeli Security Forces. Its purpose is to "subvert the typical gaming stereotype of Arabs as bad guys…" *The Washington Post* reported several controversial Islamogames (including several there were funded by Hezbollah) that have made their debuts since the rise of global terrorism (Vargas, 2006).

Conclusion: A video game's contents (with any overt or covert messages) are likely to be controlled by the funding agencies, ranging from the video game publishers (corporations), to sponsors, to independent game author(s). The issue then becomes who is 'teaching' the player, and to do what? While older gamers may be able to discern between good and evil, and right from wrong, children whose mental faculties are not yet fully developed, may not be the capability to do so.

8.3 The Rising Controversy

Amidst all the good reviews and fanfare about video games' potentials for learning, one issue has remained a hot button topic that is strongly debated in various public (online as well as political) forums. After several school shooting tragedies in recently years, the amount of graphical violence depicted in video games quickly drew national attention. Reportedly, some of the teenage shooters were found to have used violent games (e.g., *Doom*, and *Wolfenstein 3D*) as practice for their crimes. Based on the gamers' demography report (ESA, 2006), 70% of U.S. households and 92% of U.S. children below 18 years of age (Beck & Wade, 2004) play video games every week. Certainly, a good percentage of teenage male gamers would have access to 'classic' violent games (such as *Doom*) at one time or another. However, access alone does not a killer make, and such an implication upsets many avid gamers. When a gamer posted a weblog comparing the rise of video games against decreasing U.S. crime rates since 1993 (Ferris, 2005), it quickly became circulated as the 'research evidence' needed to dispel the association of violent youths and video games. Anderson (2003) pointed out the shortcomings of the hypothesis:

> "Three assumptions must all be true for this myth to be valid: (a) exposure to violent media (including video games) is increasing; (b) youth violent crime rates are decreasing; (c) video game violence is the only (or the primary) factor contributing to societal violence. The first assumption is probably true. The second is not true, as reported by the 2001 Report of the Surgeon General on Youth Violence. The third is clearly untrue. Media violence is only one of many factors that contribute to societal violence and is certainly not the most important one. Media violence researchers have repeatedly noted this."(p. 5)

In addition, any 'reported' rate of violence is "very likely a substantial underestimate" because violence acts that occurred *behind closed doors* are not reported (Straus, Gelles, & Steinmetz, 2006, p. 33). Although the seeds of violence beget violence, they don't always produce crime.

8.3.1 The Debate about Violent Video Games

Because several states are currently seeking to ban sales of objectionable (i.e., violent & sexually explicit) video games to minors, this topic would occasionally break out into heated debates in online forums operated by video game companies. Sometimes the focus of these debates would shift from a "ban on sales of violent video games **to minors**" to a "ban on sales of violent video games" (implying everyone). While the former is debatable, the latter is usually not well received by the gamers' community. The most interesting revelation is that those who are most vocal tend to eventually identify themselves as minors who do not want to miss out on playing objectionable video games if the ban is put into place!

The Merriam-Webster online dictionary [http://www.m-w.com/dictionary/violence] defines 'violence' as: *exertion of physical force so as to injure or abuse*[8]. Effects of media violence on young children have long been a research agenda of the Adult and Children Together (ACT) Against Violence project group of the American Psychological Association (APA). In 2004, the APA issued a Congressional Testimony on Media Violence and Children, highlighting the harmful effects of violence in video games on children (McIntyre, 2004). This was soon followed by several other hearings (*Violent and explicit video games: Informing parents and protecting children*, 2006; *What's in a game?: Regulation of violent video games and the First Amendment*, 2006). A number of states in the U.S. tried to ban sales of 'violent' video games to minors, but the decision was overturned after being challenged as unconstitutional by the video games industry (Gledhill, 2005). Is banning video games to minors unconstitutional as ruled? Are video games considered speech? The answers are not so straightforward. (Interested readers are welcome to find out more and decide for themselves by visiting the National Constitution Center Website at: http://www.constitutioncenter.org/education/ForEducators/DiscussionStarters/ BanningViolentVideoGames.shtml)

Today, the issue remains a media melee, drawing heat from politicians, educators, parents, psychiatrists, policy-makers, researchers, gamers, game publishers, conservatives, liberals, etc. Without going further, I hope you can sense the looming threat: Should this continue, the polarization of viewpoints will kill VGBL before it even has a chance to be used in the classroom (or anywhere else)!

Escalating violence and terrorism in the Middle East, coupled with the rise of politically sensitive Islamogames (see Dilemma #1) has prompted the European Union (EU) to echo similar concerns about the effects of violent video games on children, with possible regulatory measures (Bangeman, 2006). Although, to many psychologists, the scientific debate about the effects of media violence is already over (Anderson, 2004), some proponents of VGBL remained skeptical and have declared this issue to be over-rated.[9] While it is commendable that they have chosen to forge ahead in VGBL by focusing on the positives instead of the negatives,

---

[8] Definitions that are irrelevant to this discussion have been left out intentionally.

[9] I have chosen not to list any name or reference for this claim. However, it should be easy enough to identify this group of advocates. One of the telltale sign is: they tend to pioneer the use of controversial (e.g., violent) video games in VGBL research (and in schools).

is it a wise decision to brush aside several decades' worth of media violence research as trivial?

Perhaps one can get away from addressing the issue in a higher education institution, or a postmodernist community comprised of 'adults,' who can make up their minds about the merits of violence in video games. But if you are an educator, or a K-12 teacher, the contents — more importantly, the message — of video games should become your concern. Anderson (2004) finds it alarming that society would choose to ignore the risk of video games (with a greater *'effect size'*)[10], and instead, go to great length to put in place extensive steps and expensive measures to educate the public about other health issues (with smaller *'effect size'*) — e.g., HIV infections, passive smoke effects, and bone mass loss in old age. This brings us to the second Dilemma in VGBL: **How do we deal with violent games?**

DILEMMA #2: How do we deal with violent games?

Educators would do well to remain open-minded about violence in video games because there are plenty of grey areas. While blood and gore is considered to be gruesome, I venture to say that 'violence' is at best, neutral; and at worse, ambiguous. Violence *per se* is **neutral** because it is often a survival feature in nature (consider the food-chain, or hunting for food). The very notion of safety and protection indicate the existence of violence (nature or man-made). Violence is also **ambiguous** because it is loaded with moral implications. Any attempt to define what counts as 'violence' commits a person to accept some of his/her own acts as violent or not violent. Video games figure into that reflection process.

Example: Is violence 'bad'? If someone intend to harm you (with violence), are your actions to defend yourself (violence in return) justified? When United Airlines Flight 93 was hijacked, should the passengers have avoided violence and allowed the hijackers to drive the plane to its intended destination?

Conclusion: It is nearly impossible to grow up in a gamer generation without having played some 'violent' games. Some reports claim that as high as 80-90% of video game depicts some amount of violence (the percentage greatly varies depending on which report and whose definition you accept). The rating assigned by the Entertainment Software Rating Board (ESRB) is no litmus test, either. Should you hold the view that violence is evil, a 'violent' video game that upholds justice and promotes good may be considered *less evil* than one that 'forces' the players to take on the role of villains. More so, because children tend to imagine themselves as the heroes (typically the central character) in the story, leaving them with no alternative to the central character (as in the case of *Grand Theft Auto*) will result in these children casting themselves as gangsters, and worse, interpreting the acts depicted in the story as heroic things to do. The same argument works for

---

[10] Effect size of violent video games: 0.26.

video games depicting heroes as the central character, but we can all use more heroes.

Alternatives: While I recognize the fact that you need two sides (good and evil) to create tension in making an interesting story, we can either: (a) let the computer generate the bad guys as non-player characters, or (b) allow the player dual (maybe even multiple) perspectives by allowing player the choice to choose from both the good and/or the bad side (c.f., *Star Wars: Knights of the Old Republic (KotOR)* series). In that way, a more 'balanced' (albeit artificial) view may be presented, bearing in mind that the players (audience) are always subjected to the mercy of the story tellers to invoke fear, pain, horror, and disgust, as well as justice, joy and dignity.

8.3.2 'R' is for...

When a product is found to be defective because it reportedly caused harm or introduced a safety concern, the company (or industry) may issue a product 'recall' to safeguard their consumers' interests as well as protect the reputation of the company. Though such a measure would be a very costly affair for the company, it is often deemed to be goodwill and is necessary for the image of the brand and company. The company would gain consumer confidence by being **socially responsible**. Well-known cases of product recall include the 1982 Tylenol recall (the immediate action taken by Johnson & Johnson was highly praised) (Wolnik, Fricke, Bonnin, Gaston, & Satzger, 1984), the 2004 Vioxx recall (Karha & Topol, 2004), and the 2006 global laptop battery recall by several major laptop manufacturers (Associated Press, 2006).

One prominent example involving the recall of a video game by its publisher was the game *Grand Theft Auto: San Andreas*. In this game, an explicit sexual mini-game (though allegedly abandoned early in development) was left on the CD. One innovative gamer managed to 'unlock' the mini-game by modifying one 'bit' of code in the game, and then releasing the modified game content (i.e., a game *mod*) as the *Hot Coffee mod* on the Internet. The news spread like wild fire, and gamers (including teenagers or younger) soon gained access to the explicitly sexual mini-game via online sources. Despite claims of ignorance from the publisher, the ESRB ruled the presence of hidden content to be corporate negligence and changed the rating to 'AO' (Adults Only)[11]. The new rating forced major retailers like Wal-Mart and Toys-R-Us to pull the game from their shelves. The game publisher was eventually forced to recall the game and reissue it without the mini-game to revert the rating back to 'M'(ature)[12] (Thorsen, 2005).

---

[11] Adults Only: for persons aged 18 and older. The ESRB Rating symbol reads "Adults Only 18+."

[12] Mature: for persons age 17 and older. The ESRB Rating symbol reads "Mature 17+."

Although the recall was 'mandatory,' gamers who owned the original CD viewed it as a collector's item, and refused to exchange it for a new CD. Some gamers even make a good profit selling the original game media on eBay, which fetch an even higher price tag than the new game (without the mini-game). Some questions remained unanswered: Should the recall of such games be mandatory by federal law? Is the possession (and resale) of a recalled product a misdemeanor or felony?

Another video game, *Oblivion: The Elder Scroll IV*, also suffered a rating change from T(een) to M(ature) after ERSB found additional violent content and 'nude' skins that would alter the graphics of non-player characters (NPC) to include partial nudity. About six months after release of *Oblivion*, enthusiastic modders had created various skin patches to 'enable' the game to display full male and female nudity.

Some controversial contents may, in fact, not be the fault of the publisher, such as the case for *Oblivion: The Elder Scrolls IV (TESIV)*, and *Neverwinter Nights (NWN)*. After the games were released, some gamers who have extensive knowledge of 3D modeling wanted to 'improve' the appearance of certain female non-player characters — i.e., render them topless (Sinclair, 2006). Soon after, both male and female nude 'skins' began making their ways to the Internet and were distributed as free downloads to any gamers who wanted them. There is no news from the ESRB about a re-rating of *Oblivion* (yet?), although it remains a possibility.

Game mods and nude skins such as the one described here were *presumably* targeted at adult gamers, and were created by gamers, for fellow gamers. Though this outcome is not directly the fault of the publishers, it does not negate the fact that minors can gain access to these free nude patches through public download. Requests of nude skins for role-playing games have become increasingly common. Because *mods* were distributed via the Internet, there is no viable way to curb distribution at this moment.

Despite video games being a newcomer in (mass) media, much controversy has been associated with their content, and much has also been written.[13] The ESRB was established in 1994 as a move by the industry to 'self-regulate' rather than waiting for government to impose censorship. Yet, many questions remain unclear: Are the measures of self-regulation sufficient? Are the regulatory measures enforceable? Is the ESRB too harsh or not harsh enough in rating the video games?

Studies conducted by the Federal Trade Commission (FTC, 2006b) have shown that children can indeed purchase video games beyond the approved rating in stores without the accompaniment of an adult, or ID check. Under-age children have been known to go to gamers' (online) forums to inquire where to get access to such games, and user's advice typically includes gaining access through an

---

[13] The Wikipedia contains a detailed write-up, albeit from a video game proponent's point of view. Available online: http://en.wikipedia.org/wiki/Video_game_controversy

older friend or direct purchase through the Internet. Hence, even though the ESRB does a relatively good job in rating video games, it can do little to safeguard or prevent children from gaining access to questionable content. Besides, whose responsibility is it to enforce under-age purchases? Such questions have no doubt prompted the introduction of the Family Entertainment Protection Act, where approval from Congress is sought to allow the FTC to:

1. Conduct and publicize the results of an annual secret audit of businesses to determine how frequently minors who attempt to purchase video games with a Mature, Adults-Only, or Rating Pending rating are able to do so successfully; and
2. Conduct an investigation into embedded content in video games that can be accessed through a keystroke combination, pass-code, or other technological means to estimate certain data about video games with embedded content. (Clinton, 2005)

How will the Act (if passed into law) affect VGBL and schools? Will the possession of M(ature) or AO (Adults Only) games by schools or school children be illegal and punishable? We have reached the third Dilemma in VGBL: **What's Right?**

DILEMMA #3: What's Right?

According to the Merriam-Webster online dictionary, 'anarchy' is defined as: (a) a Utopian society of individuals who enjoy complete freedom without government, and (b) absence or denial of any authority or established order; and 'anarchism' is "a political theory holding all forms of governmental authority to be unnecessary and undesirable and advocating a society based on voluntary cooperation and free association of individuals and groups." [Retrieved January 24, 2007, from http://www.m-w.com/"]

The *gamescape* within video games is an anarchistic new world that allows an individual to do whatever he or she wishes, thus offering "complete freedom without government." In comparison, the real world simply offers abundant restriction and rules. For children, the world of video game is one without parents' nagging, and one where they reign supreme! There is no law in the gamescape, except the physical laws built in by programmers. There are no moral values whatsoever, except the views and opinions of the game writer(s).

Example: If you obey the laws of the game world, you will do well and reach the happy ending as intended; if you challenge the values of the game world, then you will likely lose or have a bad time. For example, in KoTOR, you learn the code of Jedi, and how to progress in the game from a Jedi Apprentice to a Master. If you disobey the Jedi code, or ignore their teaching willfully, you will not succeed in your training, nor arrive at the ideal ending. In another word, you have turned to the 'dark side.'

Conclusion: If one of the strongest teaching points of the video games is immediate feedback, who dictates what types of reinforcement the player gets, and at what time? Proponents of video games have argued that this is the main reason for VGBL, but from the perspective of the game designers, that would be the last thing on their minds (see Hopson, 2006). A good game designer's task is to 'hook' players with the game and to draw them deeper and deeper into the 'story' relentlessly, so that when they finally stop, the tension is released as a 'high.' This is not unlike the dark art of 'getting high' through drugs, just much more sophisticated. For now, the reinforcements are there solely to teach and ensure correct 'button meshing' for the execution of 'special kills.' There is no other consciousness (if I may put it that way) in the game, except the collective mind of the story tellers/programmers/script writers (i.e., the programmed instruction). Successful video games that have sold by the millions of copies have achieved their purpose: successfully reinforcing the gameplay mechanism in the minds and muscles of millions of players, regardless of their upbringing, socio-economic status, genetic make-ups, or moral values. This is why correlation studies trying to link violent crimes with video games appear to confirm the link, yet cannot satisfactorily explain why millions of players do not turn into killers overnight (c.f., similar scenarios were explored in *Resident Evil*, and *Dead Rising*).

8.3.2.1 Moral Development of Children

C. S. Lewis believed that man cannot escape making moral judgments because "every action presupposes a goal toward which the actor acts, and the goal (no matter how clinically it is expressed) represents a judgment of value" (West, 1996). Hence, we all make choices — right/wrong, good/evil — based on our scale of moral reasoning. In this information age, moral development is of particular importance because the advent of Internet, online chatting, file-sharing, and video games are giving rise to rampant hacking, software piracy, copyright infringements, irresponsible speech, and other 'creative' *immoral* possibilities (Willard, 1997). Without a strong moral foundation, many young people will fall prey to the snare of immoral decisions without even recognizing the dangers of crossing the line: "I am not hurting anyone by downloading these files..." As researchers, educators, and parents, our moral duty is to help children develop acceptable moral values likely to help them function as *socially responsible* citizens of the nation and world. I will attempt to discuss this by restricting the discussion below to just two samplings of moral development theories: (1) social domain theory, a modern view, and (2) moral development theory, a classical view.

8.3.2.1.1 Social Domain Theory

Social domain theory (Smetana, 1999) suggests that a child constructs his/her social knowledge (including morality) through social interactions with others, including parents, teachers, and peers. Parents play an important role in the child's moral development, because they can facilitate that process through the mechanism of affective parent-child interaction. Hence, many psychologists advocate parent involvement to co-view television and co-play video games with their children. However, this is not always possible. One important factor could be the omission of a public education about risks associated with currently available Commercial Off-The-Shelves (COTS) video games (Anderson, 2004). Many parents simply do not know what their children are playing behind closed doors or at an older friend's house. One nationwide study revealed that 68% of underage teenagers were able to purchase M(ature) rated video games from retailers (FTC, 2004). Another study showed some 12% of children (3rd through 12th grades) reported they willfully played video games (at home and elsewhere) that they knew their parents would not have approved.

8.3.2.1.2 Stages in Moral Development

Kohlberg's moral development theory (1981) is not based on maturation and social development, but is based on people's levels of thinking and solving a moral dilemma; that is, a person's moral viewpoint will expand (progress) when he or she engages in discussion and debates with people around him/her. It is considered very difficult to 'jump' stages because each successive stage builds upon the moral reasoning of earlier stages. Kohlberg did not think people would regress in their progress through the stages. His model of moral development may be presented as six consecutive and progressive stages as shown in Table 8.1.

Table 8.1 Stages of Moral Development According to Kohlberg (1981)

Exhibit by	Child	Adolescent	Adult
Level 1 (Pre-Conventional)			
1. Obedience and punishment orientation	✓	✓	✓
2. Self-interest orientation	✓	✓	✓
(What's in it for me?)			
Level 2 (Conventional)			
3. Interpersonal accord and conformity		✓	✓
(The good boy/good girl attitude)			
4. Authority and social-order maintaining orientation		✓	✓
(Law and order morality)			
Level 3 (Post-Conventional)			
5. Social contract orientation			✓
6. Universal ethical principles			✓
(Principled conscience)			

Typically, the *pre-conventional* level of moral reasoning is exhibited by children, the *conventional* level of reasoning by adolescents and the *post-conventional* level by adults. (It is possible to find adults who exhibit lower level reasoning, if they did not progress very far in moral development.) In Stage 1, the child assumes everyone else's moral reasoning is the same as his or hers. Furthermore, the degree of 'wrong' is directly related to the severity of the 'punishment' received; hence, if someone gets away without punishment consistently, the interpretation would be the action is morally acceptable. In Stage 2, the child is interested in meeting the needs of others so long as there is something in it for him/her. Hence: "What's in it for me?" In Stage 3, the adolescent recognizes social roles and tries to fit into those to meet other people's expectations (to be a good boy/girl) and to remain in good social standing. In stage 4, the adolescent or adult will feel the increasing need to obey laws and uphold society conventions to ensure the society functions well. In Stage 5, the person values human options and human rights, and recognizes that laws are social contracts (the basis of democracy). Stage 6 remains a theoretical stage based upon universal ethical principles and abstract reasoning.

Having described the six stages of Kohlberg's moral development model, what implications may one draw regarding children playing violent (or objectionable) video games? We have found the 4th Dilemma of VGBL: **Can video games affect the moral development of children?**

DILEMMA #4: Can Video Games Affect the Moral Development of Children?

Children's moral development is dependent on social interactions and also in the type of moral dilemmas they must experience and resolve. The best way to facilitate moral reasoning is for parents and teachers or other authority figures to guide and model acceptable and expected moral behaviors. When a child is allowed to remain in the game world for a long period of time, his or her moral reasoning will be shaped by the moral values of characters encountered within the game (including the storywriter, the gamers' constant NPC companions, or the online game guilds). Many player characters will be evil characters meant to advance the storyline, but that inadvertently exert influence on the moral development of children.

Examples:
Stage 1: For children in this level of moral reasoning, morality is tied to the amount and severity of punishment received. So if a gangster should get away from punishment when committing a crime, then it must not be morality wrong, or at least must be morally permissible. Perhaps the hidden message is "you can do it, just don't get caught!" If these children should get their hands on such video games (and we know they can), what are now considered as heinous crimes (in the eyes of society) will no longer have the same 'heinousness' to them among those

in younger generation because they will have been raised with a different set of moral values than their elders: namely parents, teachers, and other authority figures.

Stage 2: Children at level 2 of Kohlberg's moral reasoning are also at risk because of "what's in it for me?" thinking. For example, if a video game rewards a player with 500 experience points for stealing a car, and an additional 500 points for killing the policeman who come after him/her, the player will go for the combination of stealing the car *and* killing the policeman to get maximum experience points and reach the higher game levels faster. Hence, they are rewarded for making a morally evil decision.

Stage 3: Adolescents in this stage of Kohlberg's scheme may appear normal, do their homework on time, fulfill their duties, and behave well (i.e., be good boys/good girls) so that they can get to their video games (often set up as reward systems by parents for good behaviors). It's just like playing video game: if I do these steps, I can get those 500 points! Because social conformity works both ways, they will also try to be part of the 'in-crowd' and play whatever video games their best friends are playing. With increasing availability of network-games and Massive Multiplayer Online Games (MMOGs), the online guild (or pack) of players tend to influence one another with whatever moral reasoning they have attained collectively so far. Some of the game guilds are not unlike mobs, as they often organize 'raids' and take great pride in annihilating other players' guilds.

Stage 4: Adolescents in stage 4 of Kohlberg's model have an increasing need to uphold laws and obey societal conventions. However, if they have grown up playing video games, they would have about a decade worth of training (recall the '10-year rule' in making a chess grandmaster). By now, they would think and behave as citizens to the *game world*, and see the real world as a *meta-gamescape*. Who is to say which world is more real? (The scenario has been depicted in the *Matrix* trilogy). If gamers spend more than half of their waking hours in the gamescape, will they not lose themselves in it? (Many soldiers who have returned from a war zone after spending many months there experienced a similar social adjustment problem. School shooting incidents at Columbine and Virginia Tech similarly revealed student-gunmen who had allegedly practiced the "shooting rampage" using virtual worlds first, before crossing over to "reality.") The laws and societal conventions they know best are the ones they lived — the ones presented in their games.

Conclusion: Game developers have moral and social responsibilities to see that their video game worlds uphold the same laws and social values as the real world. It is not the violence in games that are questionable, but the subliminal moral values (good vs. evil) being taught in association with the violence.

8.4 In A Galaxy Far, Far, Away...

Evil, to some people, is the root of violence. To others, it is something or someone that brings sorrow, distress, calamity, suffering, misfortune, or wrongdoing. There are also those who believe evil to be a cosmic driving force that gives rise to various evildoings in society — much like the 'dark side' of the force depicted in the Star Wars universe.

Are young children susceptible to the forces of evil? Take Anakin Skywalker of the *Star Wars Prequel Trilogy*[14] (Lucas, 1999, 2002, 2005) for example. He seemed to have it all: brilliant, sharp, adventurous, passionate, even identified to be the Chosen One. And yet, what happened to him? Young Anakin did become Darth Vader. Over the years, the changes in his thinking process were so gradual that it was almost imperceptible; and no one around him knew what to look for, or where to begin.

The dark side is dangerous because it corrupts whatever it touches. In the Prequel, the Sith Lord (Darth Sidious), who represented the ultimate embodiment of the Dark Side, was able to prevail against the Jedi Council because:

1. he is exceedingly resourceful and powerful;
2. he was knowledgeable — he did his research and understood his opponents;
3. he was a deceiver — through the guise of Palpatine, Darth Sidious was able to conceal his existence from the Jedi Council. He also pretended to be on the verge of death when fighting the Jedi Master, Windu, and incited Anakin to help him;
4. he knew where and how to strike — Darth Sidious manipulated Anakin's emotion for Padmé to its evil end,
5. people who were working for the dark side might not realize they were being used — Count Duku did not think he was expendable, and the Separatists thought they had found an ally;
6. weaker beings might not even realize the insidious scheme — the Clone War was part of a bigger, insidious plan; and
7. the dark 'power' is extremely addictive — after Anakin tried it a few times, it became very difficult to break free.

Further, by the time the dark side reveals itself, it is often too late: The Jedi Temple has been torched, and all the younglings are dead. Who might the Sith Lords be in our 'universe'? While their identities are likely to be well-hidden, one telltale sign is that they tend to target the weak: the children.

---

[14] Star Wars: Episode I, II, and III are commonly known as the Prequel Trilogy, while Star Wars: Episode IV, V, VI are known as the Original Trilogy.

8.4.1 Marketing to Children

Because children's mental faculties have yet to become fully developed, they lack the ability to decide or control their course of actions (Wabitsch, 2006). The onus thus falls on the parents and other adults around them to provide the necessary advice, example, teaching, and life-style that will groom the child into a socially responsible adult. Speaking against the marketing of unhealthy foods to children (that resulted in child obesity), McGinnis and colleagues (2006) underscore the need for greater social responsibility: "Creating an environment in which children in the United States grow up healthy should be a high priority for the nation. Yet, the prevailing pattern of food and beverage marketing to children in America represents, at best, a missed opportunity, and at worst, a direct threat to the health prospects of the next generation" (p. 1). The same can be said about the marketing of unhealthy games to children, and the need for social responsibility to create a protected environment for a healthy development of moral values in our children.

According to the American Psychological Association (APA), the advertising industry spent some 12 billion dollars yearly marketing to children through advertisements placed on television and the Internet (Dittman, 2004). It should be no surprise that the advertising industry's latest stomping ground is to be video games (Mehta, 2003; Wong, 2005). Their intention is to turn every object within a video game into a product (i.e. *'productization'*), such as apparels, soda cans, and billboards, to be used for branding and advertisement. Because advertising to children is so effective and essential to survival of the advertising industry, some marketing companies have learned to tap into child psychology principles to fine-tune their marketing strategies. Others simply hire a professional psychologist as their consumer market consultant to do the job more effectively. While there was an outcry from some psychologists who believed this practice to be unethical because it may cause psychological wounds to children (Clay, 2000), others are happy to work with the advertising firms because there is good money to be made. ("What's in it for me?") This prompted the APA to set up a task force to look into the matter (a report of the task force is available online[15]). However, just like what happened in the attempt to regulate sales of violent video games to minors, any suggestions to regulate the ethics of fellow psychologists has quickly been labeled as unconstitutional.

8.4.1.1 Fast Foods, Video Games

Massive campaigns to advertise to children can also be found in other industries, such as fast foods. Perhaps by examining the events that are happening in the fast food industry we can foresee what may eventually take place in the video games

---

[15] http://www.apa.org/monitor/jun04/apatask.html

industry. Somewhat similar to video games, fast food companies have likewise associated their products with *'fun'* ('happy' meals). Despite medical studies that consistently demonstrate strong associations between fast food consumption and obesity in children (Bowman, Gortmaker, Ebbeling, Pereira, & Ludwig, 2003), the fast food industry has always maintained a position that the industry is blameless and that their products may not contribute to obesity at all (Brownell, 2004). Brownell pointed out that the fast food companies then "cast themselves as victims of food activism and decry attempts to curb business as usual." Such response only led to increased talks about regulating advertisement of fast food to children; and in extreme cases, the complete banning of fast food sales in certain towns: The banning of the use of trans-fatty oil in fast food chain in the state of New York is one such example. All of these incidents reminiscent of the uprising in protests against sales of unhealthy video games to minors.

The FTC and the Department of Health and Human Services (DHHS) finally issued a joint report (2006a), urging the food industry to adopt responsible marketing strategies to help fight childhood obesity. While they recognized that childhood obesity may be caused by many factors; the FTC and DHHS also believed responsible marketing would play a positive role in improving children's diets and exercise behavior. The FTC also indicated that it would monitor industry efforts more closely (FTC, 2006a), although it remains to be seen if the fast food industry will mend their ways.

One reporter claimed (Donahue, 2001), "They make this junk because people buy it." Whereas a more sympathetic Schlosser (2001) describes the *dark side* of the fast food industry in the following manner: "The executives who run the _fast food_ industry are not bad men. They are businessmen. They will sell _free-range, organic, grass-fed hamburgers_ if you demand it. They will sell **whatever** sells at a profit" (p. 269). The same quote can be applied to video games with a few modifications: "The executives who run the _video games_ industry are not bad men. They are businessmen. They will sell _wholesome, educational, and morally good video games_ if you demand it. They will sell **whatever** sells at a profit." [*italics* mine] One has to wonder, will gamers ever make the demand for wholesome games?

8.4.1.2 They Make This Junk…

If you are wondering about the connections between the fast food industry and the video games industry, consider this: Many video games are both fun and good. Some are suitable for people of all ages, while others are targeted towards older teens, and even adults. And yes, some video games are just plain junk! Furthermore, the video game industry will continue to produce this junk if people demand it, and continue to buy it. Every game in the Grand Theft Auto (GTA) series turned out to be a best-seller; there were more than enough demands shown and, of course, money was made. In another study, Hanninger, Ryan, and Thompson, (2004) found video games to overwhelmingly reward players for destructive behaviors: destroying objects (46%), injuring non-player characters (NPCs) (90%), and killing (69%). Some 54% of the games depicts Xenocide (or killing of aliens); and 63%, homicide. No wonder researchers were alarmed!

Massive Multiplayer Online Games (MMOGs), such as Sims, were once considered an innovative game genre that promotes social interactions. Because many MMOGs allow gamers the freedom to customize their online appearance by species, race, color, gender, shape and size; this enables gamers in holding different perspectives in gameplay as well as the freedom to align with one another to form social alliances, such as factions, or guilds. In theory, such flexibility in perspective ought to promote open-mindedness, patience, and tolerance among players. Instead, because MMOGs encouraged player-versus-player (PvP) activity, this has resulted in warring guilds organizing killing raids in attempts to annihilate one another. Gamers who condemn warring activity in the world today have little moral defense in supporting and encouraging the sales of ultra-violent video games that promote genocide, or worse, targeting their sale to children.

Many industries are increasingly looking to pay for in-game advertisement (Associated Press, 2007) to promote their products. It may be a matter of time before some industries decide to pay for the production of a 'serious game' to teach children about certain ideologies, such as the value of 'happy meals,' or 'diet Pepsi,' or pizza, or killing infidels? Right now, in-game advertisements already include deodorant (Associated Press, 2007), musical bands (Stuart, 2007), sport shoes, cigarettes, and alcohol (beer). More will find their ways into video games because the *gamescapes* remain unregulated by the FTC. If video games should ever become protected as 'free speech' under the First Amendment (that's what the industry and retailers want), it will be very difficult to regulate the advertisements in video games because they will become part of a protected 'free speech!' This could liberate 'obscenity' and other 'technicalities' that are currently ruled to be unprotected by the First Amendment. A worse scenario would be that the advertising industry eventually become the Sith Lord behind video game media, and gamers would find more 'advertisement pop ups' than those on an AOL browser. The dark side will stop at nothing to achieve personal gain. Now, if you worked for them, what would you do?

8.5 Rating Video Games

The *'messages'* contained in the distributed media (e.g., television, Internet, and video game) have always been the foci of media controversies. Because it is possible for indiscriminant free speech to stir up revolts and social unrests, presenting a fair and balanced view has been a longstanding standard in media making in many countries. For this reason, media rating systems have been created by ways of government censorship or industrial self-regulation, with the intent to balance freedom of speech with peacekeeping, for the purpose of greater social good. It should be noted that many countries in the world have their own independent video game rating systems. Not only are some countries more stringent in rating video games than others, they will also not hesitate to ban a video game that they considered inappropriate regardless of its rating.

In the U.S., the ESRB has assume the responsibility of (self-) regulating the video games sold in the country so as to avert government censorship. The ESRB rating system actually does a remarkable job in checking for 'problematic' contents, including: animated blood, realistic blood, mutilated body parts (gore), cartoon violence, fantasy violence, intense violence, profanity and sexual references, strong language, strong lyrics in music, comic mischief, crude bathroom humor, mature humor, reference to and use of tobacco products/drug/alcohol, simulated gambling, real gambling, partial nudity, full nudity, sexual violence, sexual acts, and suggestive themes. Three categories may be directly applied to VGBL: adult assistant need (early childhood), informational, and edutainment. (More information can be found on the ESRB Web site, at http://www.esrb.org).

Even though the ESRB rating system is rather comprehensive and did a good job in rating **current** video games, it cannot be used as a litmus test for VGBL because it faces constant challenges in the grey areas — due to the lack of appropriate descriptors or categories. In the years to come, it is foreseeable that many more categories will need to be added. Because game makers continue to push against the edge of the envelope, some of the latest controversial materials have managed to slip through the cracks of an aging ESRB system. The question then comes down to: Should some of these categories be expanded?

Take same-sex relationships for example. No doubt some parents may be shocked to learn that some 'T'(een) rated video games already depicted same-sex relationships (e.g., *Bully, Jade Empire, Sims 2*), same-sex marriages and even sex slavery (e.g., *Temple of Elemental Evil*). As the debate to legalize same-sex marriage continues to rage across the country, would adding it or not adding it as a category fuel the already heated debate? Logically, there are more advantages to both camps if a separate category is created for video games containing same-sex references in that those who are interested can seek out these games, while others can avoid them. In that way, players would know what 'contents' to expect and parents would feel that they could make better informed decisions, and not be shocked by any surprises.

8.5.1 A Failing Scheme

Not all video games available in the U.S. can be found in the ESRB rating database. Because some video games have not been submitted for rating by ESRB, blame should not be placed on the Board for inappropriate materials in these cases. Nevertheless, this highlights a looming crisis because publishers of 'controversial' video games are able to circumvent ESRB entirely by releasing the video games online. While amateur and independent video game makers may have chosen this route for one reason or another, video game developers with extremist ideologies will do so deliberately and in addition, offer the game for free to increase distribution.

In one example, the developer of *JFK: Reloaded* even put together prize money to increase sales and publicity. According to Wikipedia, the winner of the "*JFK: Reload* shooting accuracy competition" allegedly received $10,000 for his or her effort! Other examples of unrated controversial games that are available online include the Islamogames discussed earlier, the *Super Columbine Massacre RPG*[16] (based on the Columbine High School massacre), and the latest *VTech Rampage*[17] (based on the Virginia Tech massacre).

These disturbing trends bring to mind the collapse of the Hays Code[18] (the first Motion Picture rating system) in 1967, when distributors blatantly ignored the then industry-wide motion picture Production Code (social contract) to release controversial films. It seems the ESRB may be in danger of a similar fate. Should such controversies continue to escalate in the coming years, it is likely that public outcry will force the government into actions to eventually impose certain regulatory measures.

According to the ESRB's Website, the organization does not currently "have the authority to enforce its ratings at the retail level."[19] It should come as no surprise that the latest 'Undercover Shop' study by FTC (2006b) revealed that 63% of children aged 13-17 were able to purchase 'M'(ature) rated video games at a local/regional store even when unaccompanied by an adult. Certainly public education about the ESRB rating and the risks of some of the video games is in order. It may be time for the ESRB to consider introducing an expanded, improved, and clearly enforceable rating system, maybe even seek help from the government for enforcement.

---

[16] "Super Columbine Massacre RPG" depicts the players as the two shooters in the Columbine school shooting. The author believed this is his way to preserve history by making the game, and distribute it freely as a 'social documentary.' The game was not submitted to ESRB for rating but was self-rated as 'M'(ature). Since its debut on the author's web site, the game has been downloaded more than 100, 000 times. It is not known how many young children have downloaded the game for play.

[17] "VTech Rampage" is an online game about the Virginia Tech shooting. It rose to controversy because the game maker posted an online note stating he would only remove the game from circulation, and even issue an apology for making it, in exchange for certain amount of "donation."

[18] In 1967, the Hays Code was finally replaced by the Motion Picture Association of America (MPAA) rating system, which is less stringent in moral code.

[19] See the section: "Do retailers support and enforce the ESRB rating system?" Available at ESRB's FAQ site: http://www.esrb.org/ratings/faq.jsp#21

8.5.2 Independent Rating System

If video games (commercially or independently made) are to have a future in VGBL within the public school classrooms, then the issue of rating inadequacies must be addressed. Another alternative would be to set up an independent VGBL rating board to evaluate and review video games for the VGBL market, based on a set of agreed upon educational criteria and/or moral guidelines. While the definition of 'morality' is also wrought with ambiguity (Gert, 2005), it is hopefully less controversial when applied in the case of public education or P-12 children in general. One independent rating board, the United States Conference of Catholic Bishops' *Office for Film and Broadcasting* (OFB), currently rates movies based on overall 'moral tone' of the films. (Note: The OFB does not rate video games at the moment.) The new rating board may be set up with the help of a consortium of organizations to create an online rating database[20] for VGBL (or serious games), complete with suggestions on how the video games may be used in a learning situation or a classroom setting.

The worse that can happen is for the game publishers to turn a deaf ear to the educators' pleas for assistance in making games that are suitable for VGBL:

> "Academic interest in games has risen quickly over the past decade, but the games industry has never shown a similar interest in academic work. Every year there are books, journals, and conferences dedicated to studying games and how people play them, but most games professionals never read this work nor attend these conferences." (Hopson, 2006)

Is it already too late? Should educators and those interested in making VGBL a reality seek other forms and sources of help?

8.5.2.1 What Kind Of Games Are We Talking About?

To this point, I have talked about video games through a monolithic approach, intentionally ignoring the many ways video games may be categorized (e.g., through game genres, game ratings, and target audiences). Proponents of video games usually chide those who condemned games for their violence or for treating video games in a monolithic manner. Truly, because there are so many different types and genres of games, not all games are bad. Yet, these proponents have tended to treat all video games (monolithically) as though they are good for learning, even proving their points by introducing controversial games in schools as if they have no adverse effects whatsoever. Some have even claimed that it is about time for evil games to be made. Truly, not *all* games are good, either! For example, it would be very difficult to use a first person shooter game (such as Halo, or Doom)

---

[20] An example is the database at Social Impact Games, available at http://www.socialimpactgames.com/

directly — without lots of modification — in a classroom for learning simply because it would become a distraction to have to shoot a few humans, or to destroy a few houses every few minutes of the lesson. Remember that without doing so, there is no hope of 'level ups' for the players (using the game's original setting)!

Of course, this does not mean that COTS games are not suitable for VGBL *at all*. Game *modding* remains a viable though time consuming way to retrofit COTS games for VGBL. In education, there have always been sporadic reports about COTS games (e.g., *Civilizations*, *Age of Mythology*, *Caesar*, and *Sims City*) that were successfully adapted for use in classroom settings. Although limited time in classrooms tended to demand that much of the game's contents must be skipped over to fit the most important parts within lesson time limits. Even though handheld games may not fit into classroom schedules at present, they can still be highly educational and fun to play after school. Examples include: *Brain Age*, *Hotel Dusk: Room 512*, and *Pogo Island* for Nintendo DS. The new Nintendo Wii console games which make use of the innovative Wii remote controller also hold great potential for VGBL.

However, for majority of the COTS video games, classroom use could be like trying to fit a square peg in a round hole; that is, they are not suitable for VGBL *'out-of-the-box.'* President of FAS, Henry Kelly, affirmed that "popular games such as Electronic Arts' *Madden NFL 2007* football title or *Tony Hawk's Underground* skate-boarding games won't help. The games would have to be created and evaluated with the goal of raising achievement" (CBC News, 2006). Clearly, researchers are interested in a new genre of video games that takes full advantage of the affordances of video games to maximize learning possibilities. Much research will be needed to find out what the new genre will be like. Those currently engaged in serious games design are likely to be the ones who may offer some insight into this. Thus far, the FAS has showcased several games developed for this purpose, including *Immune Attack*, *Discover Babylon,* and *Multi Casualty Incident Responder*.

Much has to change in the classrooms to make way for VGBL. In the years to come, one can probably expect sweeping curriculum changes to make way for larger blocks of learning time that utilizes video games. Likewise, we can also expect changes in the interactive story-telling style of the video games to include more 'real life' learning skills, and hopefully, more widely agreed upon moral values. As the industry gradually sees more game designers, there will hopeful be more opportunities for academics to work together with designers toward achieving this goal. Academic researchers who are already working on video game research are likely to report new findings in the near future that may help steer the future directions of VGBL.

8.6 Conclusion

Video games are but a tool, and as such can be used for good and evil (Gee, 2005). But with enough knowledge and motivation, it becomes all too easy for a designer (or publisher) to demonize a game and allow players to: practice flying airplane into the World Trade Center, set off Holocaust, kidnap 15 year-old boys, practice Black Slavery, re-enact heinous crimes, or engage in any other evil acts. As Prensky (2001b) pointed out, "we have seen their formidable power used for evil, it is our duty and obligation to turn these same powerful, learning tools to as many good and positive uses as possible." Instead of spending money to research how to market to human desires, advertising film and game publishers should be responsible for helping parents research means to make interesting video games that will teach good, moral, social values instead of violence and sexual promiscuity.

From a learning perspective, it would be preferably that video games created for VGBL allowed a certain degree of flexibility. It is important that video games allow not only the exploration of the physical gamescape, but also the mind and the heart. The teaching of morality is important for young children, if we want to have an orderly society. Video games should teach critical thinking skills, and present moral dilemmas so that our children can be guided through their moral development with immediate feedback from pedagogic agents such as teachers and parents. Several important features are sorely lacking in today's video games:

- **Allowing players the freedom of choice** — to overtly teach the young ones right from wrong and showing consequence as they would be in reality is highly imperative. Some games have special points which can affect dice roll results or be used for players' character improvement. Superhero games (such as *DC Heroes* and *Marvel Super-Heroes*) encourage players to behave heroically and morally by awarding special points to game actions.
- **Allowing parental control in the level of violence** — Adding a 'no-overwrite' violence level selector, or slider, should become a standard feature. The new Xbox 360, for example, comes with a parental lock that limits the amount of blood and gore based on the ESRB game ratings.
- **Reflecting the natural/societal law and order** — A great example is the game, *Fable*, where the player character's physical features will change according to his/her actions (good or evil) in the game. Such approach reflects the natural and societal law that every action will have its consequence. Even though one may argue that in a real world the actions may not be reflected in the physical features of a person, it certainly can affect what goes on inside (the heart of a man as well as the alter-ego of the player in the game).

8.6.1 Epilogue: Which Side Are You On?

In the Star Wars movies, the portrayal of 'good versus evil' (i.e., 'a light side versus a dark side') was intentionally clear-cut. Even though key characters might 'cross over' from one side to the other, viewers tended to have a good idea about which side of the force these characters embraced. As depicted in the Prequel Trilogy, young Anakin Skywalker was first introduced to the power of the dark side when he discovered strong emotions brought on by his love for Padmé and the murder of his mother. Despite the strict training of a Jedi padawan, he found it increasingly difficult to detach himself from his emotions. His close relationship with Palpatine only served to ensure his fall, as his 'mentor' made doubly sure that the youth received sufficient exposure to the dark side to be 'hooked' or 'turned.' Step-by-step, the Sith Lord tightened the noose around Anakin's neck, and squeezed out the last remnant of light in him. When he finally failed to save all whom he loved, Anakin was left with nothing but dark despair to embrace (i.e., the hatred and power of the dark side). In other words, he became a hapless pawn of the dark side of the Force. Because of the unusual timing of the series (with the 'ending' shown before the beginning), viewers already knew that Anakin would become Darth Vader, and hence they took on the role of passive observers as the storyteller relayed the fate of Anakin.

In comparison, players of the Star Wars series of video games are much more engaged. In the first few video games (e.g. *Jedi Academy*, *Jedi Knights*), players started off as padawans who were forced to overcome many obstacles in order to become Jedi Knights (i.e., a choiceless playing on the light side). Due to demands from the players, later productions would allow players to choose sides. For example, in *Star Wars Galaxy*, players may choose at the onset to sign up as a member of the Rebels or the Republic. Out of the many *Star Wars* video games, the two-part series known as *Knights of the Old Republic* (KoTOR) I and II were the most fascinating. Set in a time thousands of years before the birth of Anakin, the games were crafted in such a way that every choice made by the player would influence the light/dark force points received. The ability to choose sides is obviously much desired by many players, as both games went on to win many accolades, including "Best Game of the Year" awards. Compared with many video games of today, KotORs breadth and depth are commendable and may serve as early examples for VGBL.

Video games have become a very powerful influence in many people lives because of the number of hours they devote to playing them. Is there really a *Dark Side* to video games? What effects does this Dark Side have on young minds? Just as the Jedi Masters failed to detect the influence of the Dark Side on young Anakin until it was too late, it is possible that the same influence is at work to seduce our young people in embracing the darker side of life. However, blaming undesirable behaviors of youths on video games is just as myopic as shifting the responsibility onto parents to regulate the playing habits of youths based on game ratings. The responsibility lies with both the industry and families.

A failure to see the dark side of video games can lead to youth delinquencies, and eventually, other larger societal issues. On the other hand, the acknowledgment of the dark side may help researchers and educators to better frame their studies and issue appropriate recommendations for remediation and change. As educators, it is our calling and duty to educate and shape young minds who will, one day, lead our society. Hence, we have the responsibility to carefully evaluate and consider the effects of video games on behalf of the children, *not only* for learning but also for living moral lives. The influence of the dark side of video games deserves the attention of researchers, psychologists, educators, and policymakers.

The future of our children may well teeter on the choices we made. No doubt, the lure of the dark side is strong, but there is still good in all of us. In the end, Darth Vader chose the light side again to finally bring balance to the Force.

The choice is now yours, which side are you on?

8.7 References

Anderson, C. A. (2003). Violent video games: Myths, facts, and unanswered questions [Electronic Version]. *Psychological Science Agenda, 16.* Retrieved January 18, 2007 from http://www.apa.org/science/psa/sb-anderson.html.

Anderson, C. A. (2004). An update on the effects of playing violent video games. *Journal of Adolescence, 27*(1), 113-122.

Associated Press. (2005, September 23). A generation of game boys, girls. Retrieved September 26, 2005, from http://www.wired.com/news/culture/0,1284,68964,00.html

Associated Press. (2006). Apple to recall 1.8 million notebook batteries [Electronic Version]. *MSNBC News*. Retrieved January 20, 2007 from http://www.msnbc.msn.com/id/14500443/.

Associated Press. (2007). Ads in Video Games Get Frighteningly Creative [Electronic Version]. *FoxNews (Technology)*. Retrieved January 20, 2007 from http://www.foxnews.com/story/0,2933,242398,00.html.

Bangeman, E. (2006). EU may regulate development and sale of violent video games (Publication. Retrieved January 12, 2007, from Ars Technica: http://arstechnica.com/news.ars/post/20061215-8433.html

Beck, J. C., & Wade, M. (2004). *Got game: How the gamer generation is reshaping business forever*. Boston, MA: Harvard Business School Press

Bowman, S. A., Gortmaker, S. L., Ebbeling, C. B., Pereira, M. A., & Ludwig, D. S. (2003). Effects of fast-food consumption on energy intake and diet quality among children in a national household survey. *Pediatrics, 113*, 112-118.

Brown, E., & Cairns, P. (2004, April 24-29). *A grounded investigation of game immersion.* Paper presented at the CHI '04 extended abstracts on Human factors in computing systems Vienna, Austria

Brownell, K. D. (2004). Fast Food and Obesity in Children *Pediatrics*(113), 1.

Carroll, J. B. (1989). The Carroll model: A 25 year retrospective and prospective view. *Educational Researcher, 18*(1), 26-31.

CBC News. (2006). Video games can reshape education: U.S. scientists [Electronic Version]. Retrieved January 20, 2007 from http://www.cbc.ca/technology/story/2006/10/19/videogames-education.html?ref=rss.

Charness, N., Tuffiash, M., Krampe, R., Reingold, E., & Vasyukova, E. (2005). The role of deliberate practice in chess expertise. *Applied Cognitive Psychology, 19*(2), 151-165.

Cheng, J. (2006). Scientists call for government to help fund video game research [Electronic Version]. Retrieved January 12, 2007 from http://arstechnica.com/news.ars/post/20061017-8005.html.

Clay, R. A. (2000). Advertising to children: Is it ethical? [Electronic Version]. *Monitor on Psychology, 31*. Retrieved September from http://www.apa.org/monitor/sep00/advertising.html.

Clinton, H. R. (2005). Senators Clinton, Lieberman Announce Federal Legislation to Protect Children from Inappropriate Video Games [Electronic Version]. Retrieved January 20, 2007 from http://clinton.senate.gov/news/statements/details.cfm?id=249368.

CNN Money. (2005). What's next: a Ph.D. in video gaming? [Electronic Version]. *CNN Money*. Retrieved December 20, 2006 from http://money.cnn.com/2005/02/08/technology/ea_chair/index.htm.

Court, A. (2006, March 12). Cyber Athlete 'Fatal1ty'. *CBS '60 minutes'*

Crawford, C. (2003). *The Art of Interactive Design: A Euphonious and Illuminating Guide to Building Successful Software* San Francisco, CA: No Starch Press, Inc.

Crawford, C. (2004). *Chris Crawford on Interactive Storytelling* Berkeley, CA: New Riders Games.

Cyberathlete Professional League. (2003). Interview with Johnathan Wendel [Electronic Version]. Retrieved January 15, 2007 from http://www.thecpl.com/gamers/?p=int_jwendel.

Dittman, M. (2004). Protecting children from advertising [Electronic Version]. *Monitor on Psychology, 35*. Retrieved January 20, 2007 from http://www.apa.org/monitor/jun04/protecting.html.

Donahue, D. (2001). Read this and you won't want fries — or anything [Electronic Version]. *USA Today*. Retrieved January 20, 2007 from http://www.usatoday.com/life/books/2001-02-01-fast-food-nation.htm.

Dubner, S. J., & Levitt, S. D. (2006, May 7). A star is made: Where does talent really come from? *The New York Times*.

Ericsson, K. A., Charness, N., Feltovich, P. J., & Hoffman, R. R. (Eds.). (2006). *The Cambridge Handbook of Expertise and Expert Performance*. New York: Cambridge University Press.

Ericsson, K. A., Krampe, R. T., & Tesch-Romer, C. (1993). The role of deliberate practice in the acquisition of expert performance *Psychological Review, 100*(3), 363-406.

ESA. (2006). *Essential facts about the computer and video game industry*. Washington, DC: Entertainment Software Association.

FAS. (2006a). National Summit on Educational Games: Fact Sheet [Electronic Version]. Retrieved October 17 from http://fas.org/gamesummit/Resources/Fact%20Sheet.pdf.

FAS. (2006b). *Summit on Educational Games 2006: Harnessing the power of video games for learning*. Washington, DC: Federation of American Scientists.

Feller, B. (2006). Scientists: Video Games Can Reshape Education [Electronic Version]. *USA Today*. Retrieved January 12, 2007 from http://www.usatoday.com/tech/gaming/2006-10-17-gaming-education_x.htm.

Ferris, D. (2005). Caution: Children at play - The truth about violent video youth and video games [Electronic Version]. Retrieved January 18, 2007 from http://www.gamerevolution.com/features/violence_and_videogames.

FTC. (2004). Marketing violent entertainment to children: A report to congress [Electronic Version]. Retrieved January 20, 2007 from http://www.ftc.gov/os/2004/07/040708kidsviolencerpt.pdf.

FTC. (2006a). FTC, HHS Release Report on Food Marketing and Childhood Obesity [Electronic Version]. Retrieved January 20, 2007 from http://www.ftc.gov/opa/2006/05/childhoodobesity.htm.

FTC. (2006b). Undercover Shop Finds Decrease in Sales of M-Rated Video Games to Children [Electronic Version]. Retrieved January 23, 2007 from http://www.ftc.gov/opa/2006/03/videogameshop.htm.

Gagné, R. (1985). *The Conditions of Learning* (4th ed.). New York: Holt, Rinehart & Winston

Gee, J. P. (2005). *Why video games are good for your soul* Australia: Common Ground Publishing.

Gert, B. (Ed.) (2005) The Stanford Encyclopedia of Philosophy (Fall 2005 ed.).

Gibson, D., Aldrich, C., & Prensky, M. (Eds.). (2007). *Games and Simulation in Online Learning: Research and Development Frameworks*. Hershey, PA: Idea Group, Inc.

Gledhill, L. (2005, December 23). Judge blocks ban on sale of violent video games to minors. *San Francisco Chronicle,* pp. A-1.

Grizard, A., & Lisetti, C. (2006, June 14-19). *Generation of Facial Emotional Expressions Based on Psychological Theory.* Paper presented at the 29th annual German Conference on Artificial Intelligence: 1st workshop on Emotion and Computing, Bremen, Germany.

Hamm, S., & Carney, B. (2005). Game Boy [Electronic Version]. Retrieved January 17, 2007 from http://www.businessweek.com/magazine/content/05_41/b3954113.htm.

Haninger, K., Ryan, M. S., & Thompson, K. M. (2004). Violence in teen-rated video games. *Medscape General Medicine, 6*(1), 1.

Hopson, J. (2006). We're not listening: An open letter to academic game researchers [Electronic Version]. *Gamasutra*. Retrieved January 2, 2007 from http://www.gamasutra.com/features/20061110/hopson_01.shtml.

Javid, C. (2004). Video games promoted as effective health-care training [Electronic Version]. Retrieved January 17, 2007 from http://wistechnology.com/article.php?id=1193.

Kapur, S., & Seeman, P. (2001). Does Fast Dissociation From the Dopamine D2 Receptor Explain the Action of Atypical Antipsychotics?: A New Hypothesis *American Journal of Psychiatry 158*, 360-369.

Karha, J., & Topol, E. J. (2004). The sad story of Vioxx, and what we should learn from it. *Cleveland Clinic Journal of Medicine, 71*(12), 933-939.

Karr, C. R., Reece, D., & Franceschini, R. (1997). Synthetic soldiers [military training simulators]. *Spectrum, IEEE, 34*(3), 39-45.

King, G., & Krzywinska, T. (2003). Gamescape: Exploration and virtual presence in game-worlds [Electronic Version], 108-119 from http://www.waikato.ac.nz/film/2005papers/319B/docs/King+Kryzwinska.pdf.

Koepp, M. J., Gunn, R. N., Lawrence, A. D., Cunningham, V. J., Dagher, A., Jones, T., et al. (1998). Evidence for striatal dopamine release during a video game. *Nature, 393*, 266-268.

Kohlberg, L. (1981). *The Philosophy of Moral Development*. San Francisco, CA: Harper Collins.

Koster, R. (2005). *A theory of fun for game design*. Scottsdale, AZ: Paraglyph Press.

Lucas, G. (Writer) (1999). Star Wars: The Phantom Menace. In G. Lucas (Producer), *Star Wars*. U.S.A.: Lucas Film.

Lucas, G. (Writer) (2002). Star Wars: Attack of the Clone. In G. Lucas (Producer), *Star Wars*. U.S.A.: Lucas Film.

Lucas, G. (Writer) (2005). Star Wars: Revenge of the Sith. In G. Lucas (Producer), *Star Wars*. U.S.A.: Lucas Film.

MacMahon, C., Helsen, W. F., Starkes, J. L., & Weston, M. (2007). Decision-making skills and deliberate practice in elite association football referees. *Journal of Sports Sciences, 25*(1), 65-78.

Mager, R. F. (1997). *Preparing Instructional Objectives: A Critical Tool in the Development of Effective Instruction* (3rd ed.). Atlanta, GA: Center for Effective Performance.

Marriott, M. (2005, February 27). What makes a good surgeon? Video games. *The New York Times*.

McGinnis, J. M., Gootman, J. A., & Kraak, V. I. (Eds.). (2006). *Food Marketing to Children and Youth: Threat or Opportunity?* Washington, D.C.: National Academies Press.

McIntyre, J. J. (2004). APA Congressional Testimony on Media Violence and Children [Electronic Version]. Retrieved May 20 from http://www.apa.org/ppo/issues/mediavioltest504.html.

Mehta, S. N. (2003). Ads Invade Videogames! Savvy marketers are eyeing one of the last ad-free zones [Electronic Version]. *CNN Money*. Retrieved January 20, 2007 from http://money.cnn.com/magazines/fortune/fortune_archive/2003/05/26/343097/index.htm.

Michael, D., & Chen, S. (2006). *Serious games: Games that educate, train, and inform*. Boston, MA: Thomson Course technology PTR.

Mills, D. (2006, December). Drug Stories. *Touchstone: A Journal of Mere Christianity, 19*, 24-25.

Murray, J. H. (1998). *Hamlet on the Holodeck: The future of narrative in cyberspace* Cambridge, MA The MIT Press.

Ortutay, B. (2007). Video-game sales a record $12.5B in '06 [Electronic Version]. *Yahoo! News*. Retrieved Juanuary 12, 2007 from http://news.yahoo.com/s/ap/20070112/ap_on_hi_te/video_game_sales_3.

Osberg, K. (1997). *Spatial Cognition in the Virtual Environment* (Technical Report). Seattle: Human Interface Technology Lab.

Paleari, M., & Lisetti, C. (2006, June 14-19). *Psychologically Grounded Avatars Expressions*. Paper presented at the 29th annual German Conference on Artificial Intelligence: 1st workshop on Emotion and Computing, Bremen, Germany.

Poole, S. (2004). *Trigger Happy: Videogames and the Entertainment Revolution* New York: Arcade Publishing.

Prensky, M. (2001a). *Digital game-based learning*. New York: McGraw Hill.
Prensky, M. (2001b). Video Games and the Attack on America [Electronic Version]. Retrieved January 20, 2007 from http://www.marcprensky.com/writing/Prensky%20-%20Video%20 Games%20and%20the%20Attack%20on%20America.pdf.
Prensky, M. (2006). *"Don't Bother Me Mom – I'm Learning!"* St. Paul, MN: Paragon House Publishers
Rosser Jr., J. C., Lynch, P. J., Haskamp, L. A., Yalif, A., Gentile, D. A., & Giammaria, L. (2004). *Are Video Game Players Better at Laparoscopic Surgery?* Paper presented at the Medicine Meets Virtual Reality Conference. Retrieved January 12, 2007, from http://www.psychology.iastate.edu/FACULTY/dgentile/MMVRC_Jan_20_MediaVersion.pdf
Rosser, Jr., J. C., Lynch, P. J., Haskamp, L. A., Yalif, A., Gentile, D. A., & Giammaria, L. (2004). *Are Video Game Players Better at Laparoscopic Surgery?* Paper presented at the Medicine Meets Virtual Reality Conference. Retrieved January 12, 2007, from http://www.psychology.iastate.edu/FACULTY/dgentile/MMVRC_Jan_20_MediaVersion.pdf
Salen, K., & Zimmerman, E. (2004). *Rules of play: Game design fundamentals*. Cambridge, MA: The MIT Press.
Schlosser, E. (2001). *Fast Food Nation: The Dark Side of the All-American Meal*. Boston, NY: Houghton Mifflin Books
Simon, H. A., & Chase, W. G. (1973). Skill in chess. *American Scientist, 61*, 394-403.
Sims, V. K., & Mayer, R. E. (2002). Domain specificity of spatial expertise: The case of video game players. *Applied Cognitive Psychology, 16*(1), 97-115.
Sinclair, B. (2006). Oblivion rerated M for Mature [Electronic Version]. *GameSpot* Retrieved January 20, 2007 from http://www.gamespot.com/news/6148897.html.
Skinner, B. F. (1968). *The Technology of Teaching*. New York: Appleton-Century-Crofts.
Smetana, J. G. (1999). The role of parents in moral development: A social domain analysis *Journal of Moral Education, 28*(3), 311-321.
Smith, M. E., McEvoy, L. K., & Gevins, A. (1999). Neurophysiological indices of strategy development and skill acquisition. *Brain Research. Cognitive Brain Research., 7*(3), 389-404.
Straus, M. A., Gelles, R. J., & Steinmetz, S. K. (2006). *Behind Closed Doors: Violence in the American family*. Somerset, NJ: Transaction Publishers.
Stuart, A. (2007). Video games are the cool new music space [Electronic Version]. *Yahoo! News*. Retrieved January 23, 2007 from http://news.yahoo.com/s/afp/20070124/tc_afp/ afpentertainmentmusicgamesfrancemidem_070124021235.
Thorsen, T. (2005). San Andreas rated AO, Take-Two suspends production [Electronic Version]. *GameSpot*. Retrieved January 20, 2007 from http://www.gamespot.com/news/6129500.html.
Vargas, J. A. (2006, October 9). Way radical, Dude – Now playing: Video games with an Islamist twist. *Washington Post*, p. C01.
Violent and explicit video games: Informing parents and protecting children, Committee on Energy and Commerce, United States Senate, One Hundred Ninth Congress, Second Sess. (2006).
Wabitsch, M. (2006). *Preventing Obesity in Young Children*. Retrieved January 17, 2007, from http://www.excellence-jeunesenfants.ca/documents/WabitschANGxp.pdf.
Wallace, L. (2006). Nintendo Wii launches with midnight crowds, cheering fans [Electronic Version]. Retrieved March 1, 2007 from http://www.realmmedia.com/next_gen_news/2006/11/.

Weinand, L. (2005). 3D Stereo Technology: Is it Ready for Prime Time? [Electronic Version]. Retrieved January 18, 2007 from http://www.tomshardware.com/2005/05/02/3d_stereo_technology/.

West, J. G. (1996). C.S. Lewis and Materialism. *Religion & Liberty, 6*(6), 6-8.

What's in a game?: Regulation of violent video games and the First Amendment, Committee on the Judiciary, United States Senate, One Hundred Ninth Congress, Second Sess. 236 (2006).

Willard, N. (1997). *Moral Development in the Information Age.* Paper presented at the Families, Technology, and Education Conference. from http://ceep.crc.uiuc.edu/eecearchive/books/fte/internet/willard.pdf.

Wolnik, K. A., Fricke, F. L., Bonnin, E., Gaston, C. M., & Satzger, R. D. (1984). The Tylenol tampering incident – tracing the source. *Analytical Chemistry, 56*(3), 466A-468A, 470A, 474A.

Wong, M. (2005). Ads invade online games [Electronic Version]. *FoxNews (Technology).* Retrieved January 20, 2007 from http://www.foxnews.com/story/0,2933,174950,00.html.

Chapter 9
Blogging the Future from Multiple Perspectives:

Current Problems and Future Potentials for Educational Games

Christopher T. Miller, Christian Sebastian Loh, Katrin Becker, Luca Botturi, Michael Barbour, Kimberely Fletcher Nettleton, Atsusi Hirumi, Lloyd Rieber, and Elizabeth Simpson

Abstract

There are many issues and potentials in the field of educational games. This chapter was developed as a collaborative blogging effort among contributing authors of this book to discuss some of the current problems and the future potentials for educational games. A variety of questions were provided in a blog format to allow an evolving discussion among the authors. Topics covered within the chapter include games and the instructional design field, problems associated with games, assessment and games, cultural implications for games in education, and future potentials for games in education.

9.1 Introduction

Games in education have become a large topic of interest in the field of instructional design and education. Between numerous presentations at conferences like the Association for Educational Communications and Technology (AECT), development of gaming special interest groups, special issues of journals such as Tech Trends, and a variety of books, games in education have gained a spot in the collective conscious of educational technology. Now is a pivotal time in the area as a variety of research is beginning to emerge on the use of games in education. Over a period of a year a variety of blog questions were posted for participating authors of this book to provide their perspective on some of the problems and potentials within the area of games in education. While it is important for individuals to discuss specific issues, it is also beneficial to provide opportunities for open forums of discussion about issues relating to a specific topic.

9.2 Importance of Connecting Games and the Instructional Design Field

Over the past several years a lot of focus has been placed on games in education, particularly in the instructional design field. What do you see is the importance of connecting games and the instructional design field?

9.2.1 Atsusi

I believe it is essential that we connect the design of educational games with the field of instructional design. Over the past 4 years, I've found that entertaining game designers know little about instructional design, and instructional designers know little about game design. Like others, I'm afraid that if we leave the design of educational games to instructional designers, they may fail to realize the potentials of story, game, and play to create engaging and memorable experiences. On the other hand, if we leave the design of educational games to entertaining game designers, they may fail to apply key pedagogical principles for optimizing learning. I believe the best instructional games require a balance between entertainment and education, and that such balance may be best derived by facilitating collaboration between subject matter experts, and experts in interactive entertainment and instructional design.

9.2.2 Kimberely

Many educational games are often incredibly easy and not very stimulating. Strengthening the connection between creative game design and educational content should become a priority. The depth of learning that is explored in typical educational games is usually limited to choosing the correct fact and plugging it in. Capturing the interest of players and involving them as learners, requires that educational games have multi-faceted texture to them. Children are willing to go delve into engaging games by reading books and researching information to help them play. Many kids map out the many levels of games or spend time creating their own levels, as an extension of the game. This sort of passion comes with a well designed game, but rarely happens with games that are designed for educational purposes. If the two could be melded together, the use of games in education could really revolutionize learning.

9.2.3 Luca

I like being concrete, and I think that the real goal is having instructional designers (IDs) and game designers (GDs) talk together. This is a pre-condition to have the disciplines come closer.

I'm active both as an ID and GD, and I experience the fact that plugging content into a game can make it boring. On the other hand, all real experts in a subject matter have somehow created fun with it - so any content can be potentially suitable for a game.

When fun is at stake, I think creativity emerging from interactions between IDs and GDs is needed, much more than any "instructional game design model", which is useful only if people have learned to work together.

9.2.4 Lloyd

Following the comments of Luca, I'll go further to say that merging the areas of instructional design and educational game design would not be a fruitful path to take. I frankly do not think such a merger is possible or desirable, at least not at the micro-instructional design level (i.e. lesson design), though I see some potential in the use of game-based learning objects as a strategy for achieving the interaction-oriented aspects of the events of instruction. At the macro-instructional design level, I can see the importance of conducting a needs assessment to determine what learning outcomes should be targeted. Likewise, doing a learner analysis is important.

I suppose my pessimism stems from the fact that I view game design as akin to story writing. Although there is a process that can be taught and understood, the best stories are only written by a talented and creative writer who is allowed to follow a path that can not be reduced to a set of steps. If it were otherwise, we would all be Steinbecks and Hemingways. I'm also of the opinion that the field of instructional design has too easily dismissed the importance of the "art of instructional design" in favor of the analytic side. So, I would rather at this point allow educational game designers full latitude to explore the creative potential of this design challenge. Likewise, I think premature attempts to add an analytic element to these designs would largely quell the creative potential.

One of my doctoral students, David Noah, did his dissertation on the intersection between instructional design, story, and gaming and found, disappointingly, that even award-winning educational multimedia could not effectively merge these areas. Making seamless the game play and the learning, one of the hallmarks (IMHO) of exemplary educational game design, is very, very difficult. He found the best approaches were when the storyline was open-ended and largely under the control of the user, such as in simulations such as SimCity. His research has influenced me, however, I like the idea at this juncture of getting both groups to collaborate and co-design educational materials. The outcome, at least for the near-future, may not be an improvement in educational materials, but it might begin to foster an understanding between these design fields that would lead to new conceptions of design in the future.

9.2.5 Katrin

I'd love to be able to devise a clean, structured ID methodology for designing good educational games. Unfortunately (or fortunately, depending upon your POV) I also know something about software engineering. This is a discipline that has devoted itself to the pursuit of finding recipes for software design that do not require skilled or talented personnel. Or, to use Edsger Dijkstra's definition: "Software Engineering is programming for those who cannot." After 30+ years of trying, we still have no real evidence that our efforts are actually producing better software. I feel there are many parallels between software engineering and instructional design. Similar desires for recipes and processes we can follow that will produce reliable results.

Still the fact remains, we need to be able to design and produce good software. Ultimately, digital games are still software. True, digital games are more than software - but they are not less than software. We also need to be able to design and produce good instruction. So, our goal is to design good instructional games, when we still don't have a nice recipe for how to design good software, or good instruction, or good games. That's a tall order.

There are people now who are designing and building instructional games that look like they will be pretty cool. But, right now, most of the promising educational games are being designed by people with considerable experience, in education, software and/or games design or some combination. We can't always require that we have developers with decades of experience upon which to draw. We can't teach new people how to build instructional games by telling them to go away and acquire 20+ years of experience. SO, we have to figure out how to teach people how to do this.

In order to do that, we need to figure out how to make good games that also embody good instruction.

The point to be made in this particular context is that a significant and irrefutable aspect of designing and creating educational digital games is the design and creation of the program that is the game. I remain unconvinced that we will be able to teach people how to design educational games without also having them understand how to design digital games. Since we don't really know how to do that either our challenge is a large one.

We have other models to turn to for inspiration though. In many ways digital games share more with film and television than they do with web applications or other instructional technologies. We still don't know a formula for producing a "good" movie, although we know some of the things that are commonly found in good movies, as well as having quite a few past and present "masters" and masterpieces that we can study. We need to study the masters (which is one of the things I am doing now).

9.2.6 Luca

I like Lloyd's comment: analysis and recipes are not the point, also because, as Katrin points out, we do not really know how to design good instructional/educational games. Once more, I think we should simply go back: take board games or sports and only after you are familiar with those, move forward to digital games (just like you need to understand classroom teaching before doing eLearning).

Also, we need to stay interdisciplinary: ID cannot absorb everything, but must be able to keep in touch with the rest of the world. We need game designers exactly as we need graphic designers, because instructional designers are neither of them (usually, at least!).

Finally, I agree with Lloyd once more: games are aesthetic in nature, and are an artistic phenomenon - we need to leave more room for "artistry" in ID. It's risky, you cannot "guarantee" effectiveness and need to trust the designers, but that's the only way to think really out of the box.

9.2.7 Christopher

I believe that there should be a connection between instructional design and game design, but I also think it needs to be at the level of collaboration. I am unsure about the extent that game designers should be instructional designers or vice versa. Both groups have their strengths and they may not bleed into the other area of specialization. I think that the overall area of educational games would be strengthened if both instructional and game designers worked collaboratively through the process of game design. This would ensure that the appropriate content and sequencing is built into the game, but that the game is also engaging and connects in a meaningful way to the users so that they continue to play it.

There are many examples of educational games that tend to sway either too far to the instructional end or too far to the gaming end. In these situations you either have bored learners or engaged players that are not learning the concepts supposedly in the game. There needs to be a balance between the two and I think that is where the collaboration comes into play.

9.2.8 Sebastian

Think of the time when most computer screens were monochromatic and then the computer industry debuted monitors with 16 colors. The industries which can afford the extra premium for new monitors immediately replaced all their monochromatic monitors with the colored ones. Other industry that could not afford the premium had no choice but to wait, until the price of the 16-colors monitors became affordable … (at which time the story repeated itself with 256 colors monitors).

I believe all of us would be familiar with stories like this, where technology replacement cycle occurs at a rate where those who have, have more, and education, lag behind… The computer games industry is the "have", and they will always be in the fore front with new hardware, new technology, new 3D modeling, new UI design, new game design… The education sector (which houses the instructional designers) tends to belong to the other camp, the one that can only afford the second-hand technology. Would game designers want to listen to instructional designers, even sit down with them to ask how Games should be designed? The instructional designers are playing catch up… not the other way around. Moreover, the last time games companies asked teachers to help design games? The result was not pretty. Has the situation changed? Do instructional designers understand the game industry? (Not really.) Are instructional designers = gamers, and hence, understand how game work? (Not really.) Do instructional designers understand the mechanics of games? (Not many.) There are many instructional designers who play computer/video games…(but not enough - no critical mass).

Will game industry ask instructional designers how to design games (educational or not)? (Not unlikely.) I am not saying that it is not impossible, but it is not likely. Unless… unless instructional designers has something the game industry doesn't have… like the method to assess learning in games. IMHO, the future of serious games is assessment (Chen & Michael, 2005) not how to design games as instruction. Hence, I am thinking that instructional designers who are doing assessment of learning, product evaluation (even usability) may have the highest chance to "connect" with the game industry.

9.3 Problems with Games and Education

There have been many reported issues regarding games such as hidden content, violence, etc such Grand Theft Auto: San Andreas. While some of these issues may not be specific to games in education they can darken the perceptions people have of games and their benefits to education. What do you see as some of the problems associated with games and education? What are some ways that these problems could be resolved?

9.3.1 Katrin

Collateral learning is most often cited as a detractor by those who do not like the idea of games for education. There is grave concern over what ELSE students may learn and how teachers can CONTROL what students experience. One of the misapprehensions many teachers have is that learning in traditional settings is controlled while learning in games or on the web is not. Control seems to be equated with feelings of safety.

The truth is we don't have much actual control over ANY learning situation - influence yes, but control?? Hardly.

A cynical view might claim that the reason we APPEAR to have control in more traditional settings is that the environment is so boring and impoverished that there is little room for learning beyond what is being presented.

As for hidden content - I think that is a bit of a red herring. Non-gamers perceive a great deal of 'hidden' content because they have never played the game - much like people used to fear subliminal messages and brain-washing in early television and record album tricks (remember the brouhaha surrounding the album clues telling us Paul McCartney was dead?). To many gamers, the 'hidden' content consists of Easter eggs and other prizes provided for their amusement. The rest is really not hidden - it is part of the game play.

People fear what they do not understand.

I think another subtext that goes with these concerns is the fear that students may learn something dark, evil, or otherwise bad. One way to address this is to analyze the game being used. The use of games in education is not a simple, nor a quick fix. Learning through experience - which is what many games are about is not efficient or simple.

9.3.2 Michael

I co-taught an undergraduate course this past semester and we were fortunate enough to have Dr. Jim Gee speak to our class via a video conference. During the session, the issue of violence and undesirable content came up and the students asked Gee what he would say to an administrator or parent who expressed concerns about this. During that discussion we all agreed that in most video games the violence is not near the level of violence to which students are exposed by Hollywood films played in social studies classrooms such as Saving Private Ryan, The Patriot, or Glory - all of which are commonly shown in high school social studies classrooms.

Granted, there are critics that will still suggest that these games contain gratuitous violence and should not be promoted by educators. During our class, Gee pointed out that the vast majority of video games available on the market are nonviolent in nature and cited the Sims as an example - the best-selling video game series of all time. He also said, somewhat tongue-in-cheek, that if playing video games led young people to do the activity in the game, there would be kids planting corn all over Los Angeles because of the popular game Harvest Moon.

Finally, he went on to describe a mission from Medal of Honor, in which the player has to get to the top of the hill during the World War II U.S. invasion of Omaha Beach. "When you're finished with that level, you're shaking, as it is just an incredibly dramatic movement; and you never notice that they never actually use any blood. It just seems totally horrific," he said, comparing this portrayal to the opening scene in Saving Private Ryan, where violence and blood are portrayed graphically. In a video game, players don't have time to sit back and "watch the heads rolling down the Beach, because you'd be dead if you did," Gee said. What is engaging is the strategy behind the actions and graphics. Violence is largely the "eye candy" and gamers cannot afford to get caught up in eye candy or they'd quickly lose the game (or as Gee said during the session, "you'd be dead in eleven seconds."

Unfortunately those who do not understand the strategy, background knowledge, and skills required to successfully play these games only see the eye candy. So, I think the issue here is not what do we do about the violence and other undesirable aspects of video games, but how do we get non-gamers to understand the purposes that the different parts of the video game serve.

In the same way that someone who knows nothing about cars lift up the hood of that car and see an engine as a single unit, a mechanic lifts up the hood and sees all of the individual components. We go to the garage and say that the engine is making a funny sound, the mechanic drives the car twenty feet and says it sounds like the fan belt may be loose.

A gamer looks at a game and sees the fan belt, a non-gamer sees an engine. So how do we teach non-gamers to become mechanics of video games?

9.3.3 Kimberely

I think that educators have to be very careful to look for hidden content in all of the materials they use in their classrooms. Games are no different. It is easy to shrug off objectionable content in something and say, "It's not as bad as…", but we are in an age where educators have to critically examine what they are using in their classroom. Teachers must evaluate everything: books, websites, movies, music, and games. They are professionals who are responsible for student learning. To not address hidden content would be irresponsible.

The interesting thing about violence is that so many people have different tolerance levels for it. Cartoons, like Bugs Bunny and the Road Runner, had to have many violent scenes edited out of them. For those of us who grew up watching the more violent versions and never once thought of dropping an anvil on someone's head, the idea of editing out the violence seems extreme. Whether this made them safer for children to watch is debatable. Many children feel there is a difference between cartoons and real people. Put any six-year old boy in front of a Power Ranger program and they will be karate chopping and kicking all over the place. There is a difference in the reality factor between cartoons and live action. Currently, games are viewed much like cartoons: not as real. As video games become more and more sophisticated, however, this may change. There are those that feel that video games with violent content actually provide an outlet for aggression.

If we are using games to teach, we need to know what we are teaching. I once had a parent upset at some of the pictures used in an elementary science book. The family picture looked as if Mom was going to work and Dad was staying home. He found the hidden message offensive for his daughter to see. I had never looked at the photos in the book in that way. Although I did not agree with his criticism or mind set, I still had to defend what was being used in the classroom. No matter what tools we use in the classroom to educate children, we need to be sure it is best for the job.

As a parent I would be very angry if my children were picking up racist messages from something a teacher assigned. Educators walk a fine line, because so many people have different values of what is acceptable. In these days of school accountability, teachers need to know if they use an educational game, it will enhance the curriculum they are required to teach. Second, parents can expose their children to a violent game, but teachers have to use materials that are not offensive to many sensibilities.

Games can be wonderful in the classroom, because they provide connecting experiences for learning. Games that are enriching and have many layers for learners can be powerful.

9.3.4 Sebastian

Many people like fast food: such as hamburgers and fries. Many people like bacon. Fast Food joints (or is it Fast Food "restaurant"?) sell whatever that sells (A restaurant is packing 6 strips of bacons onto 2 beef patties, with no vegetable). Taste good? (Probably.) Popular? (You bet.) Healthy? (You must be kidding.) Cause obesity in children? (Errr let's not talk about that in the public…) Launch an investigation into Fast Food industry? (This is hate speech!)

In a parallel universe: Many children like video game meals, some video games meals are almost like a balanced meal: with Bacon, Lettuce, Tomato. They taste good, and are children approved. A very select few even have the Parents/ Teachers approval seal.

But children, being children, choose the meal by their taste. They care not about the amount of calories, nor do they read the nutritional indexes listed on the packing.

Some children can survive on video game meals alone. Some children grow fat (too bad), some become malnourished, some are bulkier, some appeared stronger, and some looked sick. Some children throw a fit if you try to remove a video game meal from their diet... Some start stealing to get the money to get a quick fix for their burger addiction... Sometimes, a child becomes so jealous of the other kids who have burger meals, so much so that he killed... Suddenly, the parents are up in arms. "Down with the video game meals!" they shout.

Is the meal at fault? Is the one who killed at fault? Is the industry that produced the meals at fault? Is the society that condones the meals at fault? Depending on your point of view, you may think it is nobody's fault... if so then the situation remains unchanged, and continues to deteriorate.

If everyone becomes a responsible citizen, then maybe there is a chance. Do all children start planting farms after playing Harvest Moon? Do all children become killers after playing Doom? Do children jump off tall buildings after watching Superman? Do children start "kicking" one another after watching Power Rangers?

While we hesitate to label all video games/movies as "bad" because of one or two bad incident, can we turn the other way and label them ALL as "healthy" and ignore the "outliers" evidence?

I am afraid if we do so, we are no less myopic.

Back to this universe: I believe the Game Ratings need a big overhaul: The "hidden agendas" need to be uncovered and rated.

9.3.5 Luca

The issue is complex, really... I'll just make 3 small points and 2 considerations.

1. I like the concept of COLLATERAL LEARNING. But watch out! A nervous and emotionally frigid history teacher can send a much worse message than any videogame just being there, even if she's a certified teacher. It's true, we never control learning because learning is part of life, and we cannot control life!
2. On the other hand, we can control videogames, and I support the idea that there should be an ethic of videogames. Game designers should be aware of that, and realize that not all tricks are usable. It's easy: sex sells, but it is not really ethical, in the sense of honorable and finally good, to use sex to sell anything.
3. I agree with Katrin and Michael - people fear what they do not know. As people get acquainted with videogames, the buzz will get clearer.

    My other 2 constructive points:

1. It's true, children do not work "I see X", "I do X". Some children work like that, and those are the lonely ones. The study of radio and TV tells clearly that the impact of media messages depends on the strength of the social environment of the receivers. if some values (love, life, health, etc.) are strong in a family that really stays together, no violent videogame can crush that. This means the big point is not videogames (violence is everywhere!) but the actual context in which education takes place. EDUCATION is the point.
2. Videogames are different from movies because in movies you SEE violence, in videogames you ACT violently. This is paramount not to discuss in vain. So it's true, videogames are not the real point. But if the social environment (families in the first place) is weak, they can actually be extremely dangerous as long as designers do not develop an ethic view.

9.4 Understanding the Benefits of Games

While games have been used as resources within the classroom the concept of games and video games as a truly powerful instructional technology resource has not been fully acknowledged. Keeping this in mind what needs to be done to increase institutional and societal understanding of the benefits of using games in education?

9.4.1 Luca

I think a first key message concerns GAMES - i.e., fun is not the opposite of learning, but is the natural reaction to a full experience. Ancient Latin said "Gaudium de veritate", that is "the joy of truth". This means re-understanding both fun (which is not drunkenness or the nirvana) and learning (which is not boring).

The second message concerns videogames, which are not only games, but also a particular kind of INTERACTIVE MEDIA. As such, they are complex and - as said discussing the previous issue - both powerful and dangerous.

Summary:

1. True fun has to do with e-learning, and so do games.
2. Videogames are a kind of games to be understood in their peculiarities.
3. Some of the problems with videogames (e.g., violence) are not a problem with other kinds of games

And for researchers:

1. When studying videogames, do not ignore games. When studying games, do not assume they are all like videogames.

9.4.2 Kimberely

Using a learning game in the classroom can be difficult. One reason why educators don't use them is because it is hard to focus learning in a game with many variables… and many states are very concerned that teachers are teaching the curriculum, not extra material. Games are considered the dessert, not the main meal in the classroom.

The time involved in learning how to play a game and then learning content from it, can take more time than a teacher with a 50 minute class period is willing to spend. In order for games to become successfully integrated into the curriculum of classrooms, games need to be designed for classrooms. They need to be engineered to take into account their purpose. Good design is essential.

Cost is another important factor to consider when using games in the classroom. Even using a popular board game can be costly, for a teacher, since several games have to be purchased in order for all students in the classroom to play at one time. Computer games are much more expensive to license. Unlike a book, which can be easily evaluated by most educators, an electronic game is much more time consuming and difficult to evaluate for content. Most games must be purchased in order to evaluate them, whereas a book store allows one to skim through a book.

One way to bring games into the mainstream of the classroom is through better marketing and evaluations. Teachers need to know how to handle malfunctions with an electronic game, they need to know the game's curriculum content, and they need to know how to help students navigate through many levels, if the need arises. As with any instructional strategy, teachers need to feel comfortable using it. A teacher who cannot explain to the public why a particular game is being used in the classroom, will not use one. This has important implications for marketing strategies of game manufacturers.

While every new teacher wants to make their lessons "Fun", in games, the word "Fun" can be exchanged for "absorbing". This is the strength of games. This is what needs to be shared at workshops, in journals, and studied through research. As students play games, they become focused on the content within the game. This is why games become effective learning tools. They allow people to become immersed in the parameters of the game. As players become absorbed in a game, learning takes place without difficulty. Games provide quick feedback and reinforcement for successful mastery.

One way to help educators understand the importance of games in education would be to include games into the pre-service teacher curriculum. The more teachers understand how powerful games can be as a learning tool, the more likely they will be mainstreamed into classrooms.

9.4.3 Katrin

There are a whole host of things to be done to increase institutional and societal understanding of the benefits of using games in education. However, as with any other technology, some people will never be convinced. But, just the same, we must still try.

Obviously, much research needs to be done using games in the classroom - we need both quantitative and qualitative studies. There are also other things that we can do.

Here are a few suggestions:

1. Showcase "good" games. It really does not help in educational circles when people keep using war games, shooters, and violent action games as examples. I am already convinced and am a big proponent of games, but even *I* get tired of hearing about GTA, Medal of Honor, etc. These may provide many players with rich, highly compelling experiences but they are not appropriate for most classrooms.
2. Help teachers, administrators, parents, etc. gain some first-hand game-playing experience. Again - I'd suggest staying away from all M-rated games, and most war games and shooters. There are SO many games out there to choose from - there were over 2500 titles released last year.
3. Create and provide teacher support (lesson plans etc.) to go with games that could be useful in the classroom.
4. Show other ways of bringing games into the classroom (as subject matter for reports, presentations, research activities, even data gathering).

9.4.4 Michael

I'm not sure if this directly addresses the question, but it is an example of a game being used as an effective instructional technology. It was sent to me be a colleague of mine:

> ————- Begin forwarded message ————-
> From: Peter Rich
> Date: Aug 14, 2007 9:50 PM
> Subject: How's this for using games in the class?
>
> The entire class is a game

http://www.boston.com/news/globe/living/articles/2007/08/14/helping_students_stay_the_course/
———- End forwarded message ———-

My question is what is he doing that others find so difficult? What barriers did he overcome and how did he overcome them? If we can answer these questions, particularly the second, I think we have the answer to your original question.

9.4.5 Sebastian

There is currently a mismatch between what commercial video games do and what serious games proponents (like us) would like to see happened in classrooms. The former aimed to continue pushing the edge of technology and make money from the process: (1) pushing for bigger and "badder" hardware, faster VGA cards and processor speeds, (2) Proprietary software (MMO client, DirectX10, vs. Mac, and so on. Case in point: Halo3 can only be played on Vista, Crysis won't even run on most high-end machines upon debut - it needs ultra high-end stuffs. The latter continue to hope for the best, while lagging further behind in terms of technology, support, hardware/software, resources…

Another level of disconnect exists between what serious games researchers would like to see happened and what educators want happened in the classrooms. We see the potentials of video games for serious use, and try to promote it, convince other, while desperately not wanting to appear as overselling games (and appeared as a less-than serious researcher)… Most educators are concerned about finishing the curriculum on time, seeing their students getting a good grade (or not failing), meeting demands from parents and administrators, and count-down to the arrival of the deserving school holidays. There are no games for them in their horizon.

The benefit of games in education is to find a way to address the differences in expectations, and to connect the dots to create the big picture.

9.5 Integrating Games into Teacher Preparation and Education

The game industry is growing quickly as a multi-billion dollar industry with computer and video games sales coming in at $9.5 billion during 2007 (ESA, 2007). This is just behind the motion picture industry with $9.63 billion in U.S. box office sales (MPAA, 2007). Beyond sales there are also many higher education programs developing that are focused on game creation and development. Games have always had a place within the classroom whether it is Heads-Up Seven-Up or a modern computer game, yet many teacher preparation programs may not actively recognize the role of games in education. If games are to be a part of teaching, how should they be integrated into teacher preparation and education? Are there are examples of how teacher preparation and education programs are addressing games in education?

9.5.1 Katrin

Teachers need a chance to play games and they need to see what's happening - most people, teachers included don't really know what's going on in serious games - most have never even heard the term. We also need to create opportunities for parents to see ad experience some of the better games. We also need to pay some attention to giving Education faculty a chance to become more familiar with the medium.

I've taught a class on game-base learning - I still have the course website up: http://www.minkhollow.ca/EdTech/DGBL/index.html and I've done several workshops at conferences as well as doing talks on why and how teachers should play (Becker, 2005, 2006, 2007a, 2007b, 2007c; Becker & Jacobsen, 2007).

9.5.2 Elizabeth

I see the problem, of the difficulty of moving games into teacher pedagogy and methodology as one of how the teacher views their role in the classroom. How they are professionally socialized into the community and culture of teaching. We currently hand down pedagogy that has been basically unchanged for the past 60 years. The current political climate has caused teachers to embrace pedagogies that lead to ultimate control over the learning and behavior of their students. The student's however, live in a different world, which encourages them to be autonomous and self-determined. We are seeing the fall out from the clash of student expectations vs. teacher control by the increasing numbers of dropouts, 88% of whom had a GPA of C or better (Bridgeland, DiIuio, & Morison, 2006)These are not the typical dropouts. These are kids who can "play the game of school" but choose not to because they do not see the relevance. The other fall out is that more and more students are being labeled as "learning disabled" by the public schools. My research shows that these "LD" students are very capable learners in the complex learning environments of videogames. In those environments they are able to read for information, communicate ideas, be effective problem solves and utilize resources appropriately. So the question is, is it the student who is disabled or the environment? Jim Gee once said to me that it was going to have to come from special education. He might be right. I think we are reaching a crisis point.

So what can we do?

1. Continue the research, if we believe that teaching with games is better for students, we must understand how and why that might be the case. For reasons too numerous to mention here, all students can and do succeed in a game environment. I don't believe we can label students as disabled learners if we have not tried to teach them in learning environments where we know they can be successful.
2. We must model how to use technology in our higher ed teacher prep classrooms and to talk about each of the components of games as they are related to learning theory, so that our pre-service students can make linkages when they are in their methods classes.
3. We must collaborate as professionals (IT, Special Ed. Content Methods). We must insist on an integrated pre-service system so that we can collaborate and work together, co-teach rather than each having one shot at the pre-service teacher as they move through the program.
4. We must work with school districts to in-service the teachers regarding the learning inherent in the video games. We have to empower teacher by scaffolding to structures the teachers are familiar with such as lesson plans and evaluation tools, and showing them how to create them around the tool, the game. We have to acknowledge the barriers the teachers face and be willing to go with them to the administration and school boards. We have to provide safe opportunities that allow us to model how work with students in immersive environments by working side by side with teachers in their classrooms.
5. We have to share with teachers how to give the control to the students (the students will be the experts on the game) and how the teacher's role is to be the content expert to bridge the students learning to the standards and to connect the student's experiences to the big picture. Games are tools, just like textbooks, it is the lack of teacher control that is the problem. We have to model for our pre-teachers how to share the "mission" with the students.

Our teacher education programs must prepare teachers to be educators who can take advantage of learning opportunities as opposed to teachers as technician and task master.

9.5.3 Luca

I think the problem that Elizabeth points out is real. I'm aware of many experiences of programs for drop-outs in Europe, and they use more "advanced" pedagogy - experienced-based, active learning, and games. So I agree, the school system is somewhat worn-out.

However, I do not think games would solve the issue - game-based pedagogy can be part of a novel "stream" in pedagogy. In order to do that:

1. Primary school teachers (at least in Europe, most of all Northern Europe) use games - and that's a start.

2. Teacher should not (only) be taught the value of games, should be "re-taught" to play and value fun as a primary experience.
3. We need to develop sound "learning games" that teachers can use

9.5.4 Michael

I think that one of two things needs to happen. The first would be that the education system in the United States needs to get past this fact fetish that it has and focus upon deeper learning. But that takes political will and getting rid of all of these silly standardized tests.

Since that probably isn't going to happen as the United States falls more and more on PISA rankings, the second thing would be for gaming companies to design games with the classroom in mind. A good example of a company doing this is Muzzy Lane Software.

Muzzy Lane is a good example for two reasons. The first is if you look at their premiere game, Making History (see http://www.muzzylane.com/ml/making_history), it is a really good strategy game that people would want to play that is also closely aligned to actual state history standards that teachers must cover.

One of the problems with this particular game is that it is still a game that takes 50-100 hours to successfully play. As Charsky and Squire noted in their dissertations dealing with the use of Civilization and teaching history, the time commitment just wasn't realistic for a classroom environment. This is another reason why Muzzy Lane Software is a good example.

When I spoke with their reps at the e-Learn (see http://www.aace.org/conf/eLearn/default.htm) conference in Quebec City in 2007, they told me that they are shifting their focus from long, involved games like Making History to games that can be successfully played in 3-5 hours. They idea behind this was to essentially bundle a bunch of thematically tied mini-games together that followed the same strategy, role-playing model that made Making History so successful. But to do so in short games that a teacher could use in their classroom effectively because they are actually able to give their students enough time to finish the game and get the full benefit of the game play.

Once game companies begin to move to a model that is more conducive to the classroom, than we'll start to see the real power of this form of learning - cause God only knows that the classroom isn't going to change anytime soon.

9.5.5 Kimberely

Teaching through games needs to become part of teacher training, but in methods courses, it is important to focus on how to use and assess students through the use of games. Pre-service teachers must develop skills in evaluating the content and effectiveness of games. Part of training good teachers is to ensure that evaluation of all materials is balanced with student needs. Games should be treated the same way. I think game companies will have to develop venues for teachers to examine and evaluate games. Using games as an instructional strategy in method courses should not be centered on learning to play one or two games, but should make pre-service teachers aware of the instructional value of games.

9.5.6 Michael

I think that many universities have some aspects of this in their curriculum. I know that when I was a pre-service teacher at Memorial University of Newfoundland, in both my social studies teaching methods courses and my effective teaching course, simulations and how they could be an effective tool in the classroom were discussed. I know at the University of Georgia the professor (Dr. Ron vanSickle) who teaches the "Economic Education in the Social Science Curriculum" course includes a paper-based simulation of how the Federal Reserve works.

Continuing with the University of Georgia, as that's a program I know fairly well, the Social Studies Education program has a graduate course in "Simulations and Role Playing in the Social Studies" that is taught by Dr. John Hoge on an irregular basis. Only a year ago now, Mark Evans (a fellow doctoral student) and myself taught an undergraduate special topics course in that same department entitled "Simulations and Gaming in Social Studies" (see http://www.coe.uga.edu/syllabus/esoc/esoc_4000_evans_sp07.pdf) for a copy of the syllabus.

So I think that the simulations (and to a lesser extent gaming) is included in some programs - and I would argue that this is more notable in the social studies because of the larger number of social studies simulations, electronic and otherwise, available - but I think that you have to look within the teaching methodology courses to find those examples.

9.5.7 Sebastian

I am doing the following within my capacity:

- Teach a number of "game courses" in pre-service teacher preparation classes (both grad and undergrad levels)
- Conduct interesting research and invite pre-service teachers to join me in exploring what video games can do and finding new insights
- Encourage my students to enter serious games competitions/showcases, such as the International Student Media Festival (ISMF) (http://ismf.net) and I/ITSEC (http://sgschallenge.com)
- Disseminate locally what games can do for schools (I asked local newspaper to come and interview school teachers who are taking my classes. Once, we were even featured in the regional evening news (TV)
- Publicize as much as possible about your classes (handing brochures and flyers to local game stores)
- Consider the venues of your publications (book chapters are good ways to reach a bigger audience than specialize journal papers)
- Include my students (pre-service teachers/in-service teachers) in these kinds of publicity as much as possible
- My children would talk about what their daddy do for a living all the time to their friends at school (stress: without any coercion on my part)

9.6 What are the Cultural Implications of Increasing the Use of Games in Education?

Games have not garnered the best portrayal in the news especially in light of violence level or hidden sexually focused mods such as in Grand Theft Auto: San Andreas. Even with the negative imagery that some have towards games they are an increasingly large part of the lives of children growing up in the 21st century. Games are also increasingly being considered as a part of the educational toolkit that teachers have, particularly with next generation games such as Dance Dance Revolution and console systems such as the Nintendo Wii that incorporate a wide variety of body movements to control the game. What are some of the cultural implications if there is an increasing use of games in education?

9.6.1 Katrin

We all know digital games have had an impact on western society, and though some vilify games as the current cause of society's ills, the truth is that we do not yet know what the impact of games actually looks like nor what its ultimate significance will be. The focus on sex and violence is a common one, and there is a very good chance that this focus obscures our ability to comprehend the real effects of games. The public (i.e. media) image of digital games tends towards the sensational and bears a striking resemblance to the kind of media coverage allotted to television when it was new as well as virtually every other new media form dating as far back as we can remember (Williams, 2005). What we are only beginning to understand is how to use this new medium of videogames for serious purposes, in other words, as a means of expression and communication. Some of this work is being done by the games industry itself which is involved in various endeavors. For example, "Square-Enix has joined with a leading Japanese textbook publisher to make teaching games, EA has licensed the Madden franchise to a coaching software company, Konami has partnered with West Virginia to rollout a DDR-based exercise program. It's not a very big leap to say every major development shop will have at least one serious game under its belt in the near future" (Sawyer, 2007).

"So this is the backdrop for the rise of social gaming: a decline in civic and shared spaces and a decline in real-world places to meet and converse with real people. As these go down, gaming goes up. Neither event is likely causing the other. Instead, here is a hypothesis about what is happening: Humans, whose need for social contact has never changed, find themselves with a desire for community and social interaction but with fewer and fewer real-world outlets. The demand for human connection has been static but stymied by the real, it has moved into the virtual. As a result, social ties have moved online as part of a virtual community trend (Rheingold, 1998). As one of the most popular online functions that bring people together, games are a particularly important site of activity to consider." (Williams, 2006, p.15)

9.6.2 Luca

Games are a particular form of human cultures, since ever. It is true, some video games (only videogames!) are under attack because of sex and violence - but the point is not the games themselves, rather the trash that they contain (as it was with movies, TV, rock-n-roll).

Using games in education in a way will make games more positive in the eyes of the society, if they can bring some advantage. As it was with sports, which have always, been part of education.

On the other hand, there is a risk: relying too much on games could make education just "fun", that is, deprive students from the ability to learn from hard work and even pain. It is true, we learn from having fun, but also from tragedies, and life is made of both. We live in a "happy thoughtless society", pushing us to play all the time. Learning to have responsibilities and to live through sufferings is also something that, as educators, we should not forget.

9.6.3 Michael

I think that each new form of media is vilified to some extent until it become either mainstream or the next new media comes along. First it was rock and roll, then television, then heavy metal and the likes of Marilyn Mason (and many others before him), then rap/hip hop, then gangster rap, and now we have video games. Many of these forms of media found their appropriate place in the classroom. Video games will follow the same pattern.

Personally, I'm much more concerned with the potential of games to change the nature of the classroom experience (i.e., that learning interaction). As Luca mentions, the possibility of school not to be a thoughtless or mindless activity, but to be an engaging experience. My fear is that games will follow the same classroom pattern as all of the other media that have been introduced, as an extension of the same old way we've always done things... Nothing more than the typical experience students have had in school for the past decade with Oregon Trail.

9.6.4 Kimberely

First, I think that we have to look at who is comfortable with playing games and make sure that the games used in education are equitable. Games will need to be used early and often in schools so both boys and girls, from many different socio-economic backgrounds develop the necessary skills to navigate through games. From the outset, attention to bias towards one gender or population must be addressed, so that culturally, we don't unintentionally create an unnecessary schism.

Michael fears that they may be an extension of the same old thing in the classroom, but I envision good educational game design as calling for more interaction between teacher and students. To have real cultural impact, games will have to make the shift from drill and practice to instruction. The role of teachers will also adapt to the media. Good teachers know their students and games and teachers cannot be used interchangeably. Educational game design will need to provide teachers with ways to manipulate the parameters of a game to address individual student needs. A new dimension to differentiated instruction will open when teachers can adjust reading times, vocabulary, font size, or complexity of ideas.

9.6.5 Sebastian

One positive sign I have observed is that increasingly game magazines and web sites have chosen to use the term "casual games" – presumably to distinguish themselves from "serious games"? I see this as a positive sign because it tells me the industry is open to the idea, even though that does not mean they will support serious games (to begin creating applications and tools to help it become a success). The industry is still driven by profit margin, and it is a very competitive market!

Another positive sign is that a number of "big guns" have shown their supports to the movement – e.g. George Lucas (edutopia.org) and Microsoft (XNA studio).

Now, if only "all of us" will begin making games (instead of continue to talk about it), perhaps we can also help change the perceptions of others… Ya?

9.7 How Can Assessment be Conducted When Using Games?

We live in an age of assessment, particularly high stakes testing. Assessment can come in many forms, but oftentimes when you think of the word assessment it brings to mind some type of test. These could be instructional tests, placement tests, and achievement tests to list a few. While testing is one form of assessment there are many other forms of assessment. If games continue to grow in popularity and use in education then assessment becomes one of the important questions to ask. How can and will assessment be conducted? Should the assessment related to using games in education be different than assessments currently look? If it is going to be different how would it look?

9.7.1 Luca

Games are holistic learning environments, that is, they tend to involve the person as a whole - you cannot play and think of something else, you wouldn't be really playing…

So, they foster competence learning in the sense developed in some European research about the Bologna process and mobility: while skill means being able to perform a task, competence means creatively dealing with some kind of situation.

Consequently, games support "real-life testing" such as running a project or simulation. But the learning that happens in game can concern smaller objectives, so I think that simpler forms should not be disregarded: a quiz can be good to test knowledge of some facts learned from a game.

9.7.2 Katrin

Modern commercial games are *already* pretty good at assessment. All we really need to do to create a solid foundation for assessment in digital games is to study how the good games do what they do. Many are competency based and players do not get to the next level until they have demonstrated a certain level of proficiency. That's not to say we can not build upon what there is - we can and should, but it would be a waste of time, knowledge and resources to ignore that which good games already do.

9.7.3 Michael

This is an interesting question, because while most games don't have the content that one would find on a high stakes test, they do help develop the skills that would make one successful in the current school environment. So there are two ways to think about this.

One is to look to see what effects games can have on students' performance in high stakes testing. This is something that we have started to do with the homemade PowerPoint games project (see the e-Learn and National Association of Laboratory Schools Symposium presentations at http://it.coe.uga.edu/wwild/pptgames/pubs.htm).

The other is the argument that is often used by social studies educators when you look at the amount of social studies being taught at the elementary and middle school levels. Most schools have pretty much cut out social studies at these levels, giving it one hour of instruction every one to three weeks depending on the area, because social studies isn't directly tested at those levels. However, if you examine the tests that are given and look at the content of the questions themselves (particularly for the English language arts exams) you'll see that much of the content of the questions are taken from the social studies disciplines (as much as 60% according to some studies).

So if you took a game like SimCity or Civilization or the Sims or Guild Wars or Grand Theft Auto and tried to find the content that actually matched up with the standards that students need to cover, then you would have a guide that teachers could use to tap into the schema that students already posses (cause let's face it, they're already playing these games, we don't need to bring them into the classroom) then you have a powerful learning tool.

I mean does it matter if the student understands the concept of opportunity cost because you used an example of the local factory or you used an example from a video game like Guild Wars or Grand Theft Auto. Isn't the main point that the student understands the concept in a way that has personal meaning to them so that they'll be able to remember it come test time?

9.7.4 Kimberely

I agree with Michael that games can provide connections to concepts that students are expected to develop for assessment. There needs to be a better understanding of what concepts are within the design of a game before teachers use it for concept development. Many questions have to be examined. How clearly are concepts explained to players? Is concept development based on prior knowledge or is it learned during playing? How long does a student need to play the game to fully understand the concept? What other concepts are students developing? How much time is needed for students to learn to manipulate and play the game? Quite frankly, if it takes too may class periods, teachers will probably choose to use something besides a game to develop a concept. If games become a mainstream instructional strategy then the gaming industry will need to provide teachers with more information so they can fill in the gaps and help students develop conceptual frameworks.

Using games as an important piece of curriculum development will change assessment. Just as assessing cooperative learning groups or performance based projects requires a different form of assessment, an assessment model for games will need to be developed. While students interact with a game, teachers will have the opportunity to individually assess student knowledge through conversations and observations. Assessment questions could be built into games to check level of understanding so that skill level does not become synonymous with learning. Instructional games could provide a better assessment model to track student learning if it is carefully developed as an integral part of the game.

9.7.5 Sebastian

Katrin was absolutely right that "modern commercial games are already pretty good at assessment." Whether the players realized or not, they are being assessed by the computer games as they progress from level to level, beating boss after boss. However, educators and school system need a "scoring/rating system" to show "how good James is compared to Sally or…" Everybody related to the education system: parents, administrators, even students, all want that piece of data. The current trend of data-driven school and education improvement through data warehouse and data mining will accentuate the situation even more.

Based on my experience as a school teacher, I want to be assured that Johnny has been "measured" according to some known learning standards should I let him play games in class. These games must be created by people who are trustworthy (to balance learning with fun in the games) and with credentials (like textbook publishers). In addition, I would really appreciate it after Johnny played the game, he would be "evaluated" immediately (just like divers and ice-skaters receiving scores after a judging). These scores can then be kept as a record of his performance. In that way, I could compare the students and spot if anyone is lagging behind and provide necessary remediation.

Using these guidelines, I formulated the conceptual design framework known as "Information Trails" (Loh, 2007; Loh, Anantachai, Byun, & Lenox, 2007). My aim is to create a serious game design methodology for others to create serious games that can be used in the classroom. The "Information Trails" methodology assesses the learning performance of players based on the decisions they made throughout the game. I am happy to report that the design framework has been completed and my development team is now working on the application for creating after-game "status reports" (a.k.a. After-Action Report (AAR) in military training).

Readers who are interested in sharing their thoughts on assessment in serious games are invited to check out the Consortium for Instructional Design, Evaluation, and Assessment Strategies (I.D.E.A.S.) in Serious Games web site (http://idt.siu.edu/ideas).

9.8 What Are the Future Potentials for Games in Education?

The idea of games has changed over the years. Early games were played with sticks and stones, in the 70s and 80s was the advent of the arcade game, and now games can be conducted online and with hundreds and even thousands of people playing at the same time. As games have increased and evolved so also have games in education, but where will they go from here? What do you see as the future potentials for games in education?

9.8.1 Michael

There are a couple of things that should come out here… The first is the evolution of games by companies like Muzzy Lane Software, that I described in my second comment to the Integrating games into teacher preparation and education prompt (see http://edgames.ed-u-tech.net/2007/10/03/integrating-games-into-teacher-preparation-and-education/). If we can start moving more in this direction, I think that the political will needed to get more games into the classroom will be there with these short games that are closely aligned to state standards.

The second is that games need to be seen as just another arrow in the teacher's quiver. Let's not forget that there are some teachers out there that are master story tellers and lecture works for them and their students because the teacher is just that engaging and that thought provoking and that entertaining. But that's not every teacher and that's not even the great story teller on every day. In the same way we find the use of games and simulations often included in teaching methodology classes in schools of teacher education, we need to remember that this is just another pedagogical strategy that a teacher can employ in the right situation, with the right tools, with the right group of students - and teachers are the best ones to make that kind of informed decision.

Finally, I think the article that Squire published in Ed Researcher in 2006 is a good place to look in response to this question. The main theme of that article, at least from what I took away from it, was that the politics of education right now isn't conducive to bringing video games into the classroom (at least not the way video games are currently constructed - the Muzzy Lane vision may serve to change that). He seemed to be advocating that those of us interested in gaming in education, particularly over the counter or commercially successful video games, focus our attention on working with youth in after school programs and using the schema that students gain from their own game play at home.

9.8.2 Katrin

"We need to consider whether we are educating children for their futures or our pasts."

Geoff Southworth 2002

In his studies of engineering education, Richard Felder found that "learning styles of most engineering students and teaching styles of most engineering professors are incompatible in several dimensions. Many or most engineering students are visual, sensing, inductive, and active, and some of the most creative students are global; most engineering education is auditory, abstract (intuitive), deductive, passive, and sequential. These mismatches lead to poor student performance, professorial frustration, and a loss to society of many potentially excellent engineers (Felder, 1988, p.680)". Just as Felder finds it appropriate to advocate for inductive teaching styles for all types of learners, it may also be appropriate to advocate for supported learner control for all. That learning is more effective and learners more amenable and responsive when they are given greater control over their learning environment is now a widely endorsed tenet. Games already do this. Control over one's environment is a key aspect of virtually all popular games, from Lord of the Rings, to Paper Mario and Metroid Prime.

Among the things successful games do effectively is teach, whether or not we as a society value what is being taught. In games, for the most part, failure is free, but allowing mistakes takes time and this is runs counter to the goal that many educators espouse, namely to make education more efficient. With budgets being continually eroded, it is hard to argue for anything in education that does not increase efficiency. Learning as defined by the first two levels of understanding in Bloom's taxonomy (Bloom, 1964), which are knowledge and comprehension, has formed the backbone of formal education since the turn of the 20th century. It is no longer sufficient. Learners today need to be able to synthesize and evaluate information and knowledge in the face of a constantly changing technological environment. The virtual worlds of massively multi-player environments and modern digital games provide diverse environments in which this approach to learning can take place, provided educational game designers are up to the challenge of designing and using games and environments that retain those qualities of games that make them compelling, while at the same time offering sound instructional interventions. One of the ways this can happen is through a balanced synergy between game design and instructional design.

9.8.3 Katrin

An issue that was brought up in the first post is one that may well be unique to the U.S. - namely the effect of the current preoccupation with standardized (and traditional) testing in light of NCLB. Many other countries either already have or are in the process of moving beyond the notion that education can be improved by rigid standards and enforced adherence. While there is an understandable call for the development of games that adhere to U.S. state standards, other countries are finding ways for games to help advance our understanding of teaching and learning rather than trying to force games to be adapted to what many consider to be an outdated model of formal education.

When it comes to "getting a education", learners have more choices now than they ever have in history. Many kids already do most of their schoolwork away from school, where they have access to their own technology and are subject to far fewer restrictions than they are in the classroom. Games offer an opportunity for us to move beyond the Taylorian factory model of formal education. If we insist on 'fitting' games into a century old model, we run the risk of re-creating the situation we had in the 80's when digital games first gained the attention of formal education resulting in the 'edutainment era'. This approach failed before; it is unlikely it will succeed now - learners will simply turn to alternative forms of education.

As I see it, our choice is to work towards moving formal education into the 21st century, or watch it become irrelevant.

9.8.4 Kimberely

Perhaps the biggest potential I see for games in the future is the way they have the potential to reach so many students at so many different levels. Games can provide the vehicle to reach students who are disenfranchised or stymied by learning disabilities. Games can provide motivation for learning. Games can change the way students see learning and themselves. I have heard very young children talk about their "Evil Minions" or "Devastating imperial Starship cruisers", based on their interaction with popular media. It's fascinating to hear and realize children really know this vocabulary. They have picked it up through contact with media. Games can provide this same collateral learning.

Games in the future could also change the ways we assess students in the classroom. In many ways, formative assessment is an inherent part of games. Players receive reinforcement as they move through levels of play. Educational games could be designed with assessment as part of the game package. A simple player survey at the beginning of the game could gather demographic information. By providing pre-game questions that collect data on reading levels, processing abilities, and reaction time, the game could adjust itself to the learner. As students play the game, the game could continuously assess information on cognitive ability. It would be possible to gather information on problem solving ability. Games could adapt to students' abilities, so that both obstacles requiring problem solving and situational success could be built into the program. Games could become more adaptable and capable of enhancing programmed learning than any learning machine dreamed up by Skinner.

Schools begin embracing games as an instructional strategy, one of two things will happen. Either games will be watered down and become generic (boring?) to meet curriculum standards or they will change the way students learn concepts. With visuals, music, and movement, students will be able to assess curriculum content in new ways that were totally unimaginable 50 years ago.

9.8.5 Sebastian

Even though Guitar Hero is a major hit, none of the "heroes" (i.e. players) would be able to play a "real guitar" (unless they also have real-life guitar learning experience using a real guitar). Virtual experience is only transferrable to real-life which occur under strict circumstances (like using a high-fidelity flight simulator to train pilots). Simplified simulators like Guitar Hero can only provide players the fun, and the illusion of success, not the actual skills.

When we create serious games for education (classroom learning), we need to bear these restriction in mind. The illusion of "success" for players is what makes games motivational (even addictive). Stick within that boundary and find an appropriate situation in education to introduce games for learning (or to create a game for that "moment") and your students will love and praise you.

On a separate note, let's not turn everything into games. In that situation, innovative ideas will turn into clichés, and game won't be fun anymore. While one can certainly make games to teach, anything under the sun (when motivated by profits), they may not be the best use of the media to teach the "content." If serious games "companies" begin running around with the video game hammer in search of nails, shoot it! We don't need a repeat of the edutainment history.

9.9 Conclusion

This chapter has focused on providing a dialogue between the contributing authors about issues related to games in education as well as some of the future potentials. While there were several issues discussed in this chapter about games in education one problem that stands out is the public image of games.

Over the past several years there have been many reports about inappropriate content, violence, and hidden content. Katrin mentions in an early posting that the issue of hidden content is a red herring, but maybe that red herring goes beyond just the hidden content. What seems to have occurred in recent history is that games have been singled out as having a variety of inappropriate content, but couldn't that be said of other media forms. How many television shows present inappropriate content and violence during the typical family evening hours of television? What about the hidden messages that used to be embedded in films to encourage trips to the snack counter? In some ways it almost seems that there is more a focus on games, because it is the newest media on the block. It appears that our society has become numb to the inappropriate content that appears in music, television, and movies, but it is something new for games. When you begin to look at the stories about inappropriate content is it really that there are a lot of cases of inappropriate content or is it many stories about a few cases that then expand the issue? Ultimately, when it comes to education, the teacher needs to appropriately select the material that will be used in the classroom whether it is video, music, or games.

This leads into another issue that was addressed in this chapter relating to the benefits of games. When educators are inundated with stories of inappropriate content in games it clouds the understanding of benefits in games. Several of the contributing authors outlined some ideas for helping people to understand the benefits of games in education. These benefits include:

1. Showcasing good games;
2. providing first-hand gaming experience to colleagues, administrators, and parents;
3. showing examples of the fun that can be connected to the learning, and
4. training individuals in the careful selection of appropriate games.

While it is important for those in P-12 education to understand the benefits of games in education, it is equally important that these benefits are understood in teacher preparation programs. Some of the ways that this could be achieved is:

1. teach about game-based learning in teacher preparation technology courses;
2. embed game-based learning within teaching methods courses;
3. make game-based learning programs available in the teaching section of libraries for student evaluation; and
4. establish graduate level game-based learning courses for graduate level teacher programs.

As the benefits of games become accepted in educational practice there will inevitably be questions about whether games can be used in assessment, particularly in an age of high-stakes testing. Concerns about assessment with games are important, but also somewhat answered in the games themselves. As Katrin describes in her posting games already conduct assessment and could even be seen as an assessment form. When playing most games an individual cannot advance a level until it has been mastered. Strikingly this sounds like what occurs in various instructional design models. What needs to be considered is how to create new ways to incorporate alternate assessment methods beyond the in-game assessment to gain a broad understanding of student mastery. Some of the ideas suggested by Kimberely are to assess through conversations, observations, and building assessment questions along with the game to gauge the level of understanding along with skill mastery. Most importantly, as Katrin established in her posting, educators need to study games and learn from the assessment models within games to develop new ideas for assessment.

Games are becoming a part of culture and as they become more mainstreamed greater acceptance will occur. That mainstreaming in fact is already occurring. Just take a look at some of the commercials for online games that feature prominent personalities participating. As the mainstreaming of games and games in education occurs, the potentials for them will grow. It is important also to keep in mind that the games will evolve as new technologies are created. As an example look at the Nintendo Wii, which incorporates a variety of body motions to control the on-screen actions.

While games will continue to evolve there are several things that need to be kept in mind. First, as Michael stated in his post, games need to be seen as just another arrow in the teacher's quiver. There are many other tools that can also be used to enhance education. Second, as Kimberely and Sebastian both discuss, educators need to exercise caution when using games in education so as to not water down the experience and thus become generic and potentially boring. Third, as Katrin describes, it is important that we explore new ways of learning with games rather than forcing it into an old model of education and thus allowing it to become irrelevant. Finally, we need to exercise caution utilizing games in education and remember one important rule: games like education have been part of human experience throughout history. And just as the games we play have evolved with changes in society and technology, so too must education respond to these forces. By exploring the potential of games in education and how games are creating their own forms of learning educators can enhance their own forms of instruction to better address the learning styles for the 21st century and beyond.

9.10 References

Becker, K. (2005, May 24-27, 2005). *Are you game? The future of learning with technology.* Paper presented at the AMTEC Conference, Calgary, Alberta.

Becker, K. (2006, November 2-4, 2006). *An hour of Play.* Paper presented at the International Conference on Teacher Education, Calgary, Alberta.

Becker, K. (2007a, June 25-29, 2007). *An afternoon of play: Introduction to video game literacy.* Paper presented at the Annual World Conference on Multimedia, Hypermedia, & Telecommunications, Vancouver, Canada.

Becker, K. (2007b, March 16, 2007). *How games can enhance learning and e-learning: There's more to video games han you think.* Paper presented at the Technology Enhanced Research Area, Calgary, Alberta.

Becker, K. (2007c, June 25-29, 2007). *Play is the beginning of knowledge.* Paper presented at the Annual World Conference on Educational Multimedia, Vancouver, Canada.

Becker, K., & Jacobsen, D. M. (2007, May 12-16, 2007). *The importance of being earnest.* Paper presented at the CADE/AMTEC Conerece, Innipeg, Maniba.

Bridgeland, J. M., DiIuio, J. J., & Morison, K. B. (2006). *The Silent Epidemic: Perspectives of high school dropouts.*: Bill and Melinda Gates Foundation.

Chen, S., & Michael, D. (2005). Proof of learning: Assessment in serious games. Retrieved July 17, 2007, from http://www.gamasutra.com/features/20051019/chen_01.shtml

ESA. (2007). Industry Facts. Retrieved October 3, 2007, from http://www.theesa.com/facts/index.asp

Loh, C. S. (2007). Designing online games assessment as "Information Trails". In D. Gibson, C. Aldrich & M. Prensky (Eds.), *Games and simulation in online learning: Research and Development Frameworks* (pp. 323-348). Hershey, PA: Idea Group, Inc.

Loh, C. S., Anantachai, A., Byun, J., & Lenox, J. (2007). *Assessing what players learned in serious games: In situ data collection, information trails, and quantitative analysis.* Paper presented at the Conference Name|. Retrieved Access Date|. from URL|.

MPAA. (2007). Research and Statistics. Retrieved October 3, 2007, from http://www.mpaa.org/researchStatistics.asp

Biographies

Editor Biography

Christopher Miller

is an Associate Professor of Education at Morehead State University in Morehead, Kentucky. He is the coordinator of the first fully online educational technology program in Kentucky. He received his Ed.D in Instruction and Administration from the University of Kentucky with a focus on instructional systems design. His research interests include web-based instructional design, games in education, classroom technology integration, and multimedia. He is also an active instructional designer, trainer, and consultant.

Author Biographies

Michael Barbour

is an Assistant Professor of Instructional Technology at Wayne State University in Detroit, Michigan. He recently completed his Ph.D. in Instructional Technology from the University of Georgia. Prior to his doctoral studies Michael was a teacher in Newfoundland and Labrador (Canada), where he served as a classroom and virtual school teacher, along with serving in a number of district-level positions. His research interests focus upon rural K-12 students learning in virtual school environments and how the process of game design assist students in developing a deeper understanding of the content material.

Katrin Becker

taught Computer Science at the University of Calgary from 1983-2006, and was responsible for many innovations in computer science education research including: methodologies for comprehensive coordination of first-year CS major's courses, application of inquiry-based and learner-centered approaches in freshman CS programs and large class settings, and methodologies for facilitating cutting-edge research with undergraduate students. She's been using digital games to teach since 1998, taught one of the first Digital Game Based Learning courses for an Education faculty, and now spends time helping others become familiar with the educational potential of games.

Her current interests include the use of computer games for learning at all levels of education, including building games in order to learn programming and other CS concepts. She holds a PhD in Educational Technology, where her work focuses on game design and on exploring methodologies for incorporating instructional design *into* rather than *onto* the game design process.

Luca Botturi

holds a Ph.D. in Communication Sciences and Instructional Design from the University of Lugano, Switzerland. He is currently instructional designer and researcher in Lugano. His research focuses on creative instructional design, design languages and games. He is active as trainer and consultant, and has founded seed, a non profit organization promoting the development of a culture of educational technologies for international development and non profit education. He is also a board game designer.

James Paul Gee

is the Mary Lou Fulton Presidential Professor of Literacy Studies at Arizona State University. He is a member of the National Academy of Education. He is the author of *Sociolinguistics and Literacies* (1990, Third Edition 2007); *An Introduction to Discourse Analysis* (1999, Second Edition 2005); *What Video Games Have to Teach Us About Learning and Literacy* (2003, Second Edition 2007); and *Situated Language and Learning* (2004) among other books. His most recent book is *Good Video Games and Good Learning: Collected Essays* (2007). Prof. Gee has published widely in journals in linguistics, psychology, the social sciences, and education.

Atsusi Hirumi

is an Associate Professor and Co-Chair of the Instructional Technology Program at the University of Central Florida. Over the past 10 years, Dr. Hirumi has concentrated on developing e-learning systems, working with universities, community colleges, K-12 school districts and businesses across the United States and in Mexico to establish online training, certificate and degree programs. His research centers on the design of alternative e-learning environments with a current focus on storytelling and game-based approaches to teaching and learning. Recent awards include the Texas Distance Learning Association award for Commitment to Excellence and Innovation and the WebCT Exemplary Online Course Award (in 2003 and 2001).

Chistian S. Loh

is Assistant Professor of Instructional Design Technology and Coordinator for the Collaboratory for Interactive Learning Research (CILR) at the Southern Illinois University Carbondale.

His research interests include assessment for and of learning with digital games, and Information Trails. He has chosen game modding using commercial game engines to create new contents for serious play and to search for new ways to integrate videogames into education, training, and classroom instruction. He teaches instructional technology and the use of open source software for online content delivery and e-Learning.

He actively promotes videogames for learning as President (2008-9) for the Multimedia Production Division (MPD) of the Association for Educational Communications and Technology (AECT), and associate editor for the International Journal of Gaming and Computer-Mediated Simulations (IJGCMS) – a new journal to be published in 2009.

Kimberely Fletcher Nettleton

is an instructor at Morehead State University (MSU). She taught in public and private schools for 17 years, serving as a principal for 4 years. She is completing her doctorate in Instructional Design and Technology, having completed her BA in elementary education at the University of Kentucky and a Masters in elementary Educational at Georgetown College, as well as a Masters in Supervision and Leadership from Morehead State University. Currently, she is the program coordinator for the Professional Development School at MSU. Her research interests are in improved interdisciplinary curriculum design as well as looking at the implications of technology applications for specific populations: special needs, gifted, gender, etc. She is currently working with Middle and High School math teachers on improving instructional strategies through the incorporation of math literacy strategies.

Dawn Rauscher

is an instructor at Flathead Valley Community College in Kalispell, Montana. While obtaining her Masters degree in Instructional Technology at the University of Georgia in 2005, she was a TA who taught pre service teachers how to integrate technology into the classroom. Currently, she teaches a similar undergraduate course at Flathead Valley Community College and she enjoys working with teachers and students in a close-knit community.

Lloyd Rieber

Lloyd Rieber is a Professor in the Department of Educational Psychology and Instructional Technology at the University of Georgia. He received his Ph.D. from the Pennsylvania State University in 1987 and is a former classroom teacher. He is interested in visualization, cognitive psychology, and constructivistic orientations to instructional design. His research focuses on using dynamic visualizations in the design of interactive learning environments. His most recent research is about the integration of computer-based microworlds, simulations, and games using play theory as the theoretical framework. He is now applying this research to support online learning environments and to help students with cognitive disabilities.

Elizabeth Simpson

is an associate professor in the College of Education, Department of Special Education at the University of Wyoming. Her research centers on examining learning outcomes using simulation video games to teach content in classrooms. Consequently, her research extends to professional development and collaboration for educators. She has over 20 years classroom experience as a Special Educator, college professor, consultant and researcher. Her research is focused on the use of video games as learning tools for business and education that will build on the strengths of today's learners and tomorrow's leaders.

Susan L. Stansberry

is an Associate Professor and the Program Director for Educational Technology and Library Media at Oklahoma State University. Her research interests include the use of media and technology in teaching and learning and organizational culture as a perspective for technology adoption. Her interest in helping teachers integrate videogames and simulations in the classroom stems from all the years she spent getting sent to the principal's office as a student, teaching and coaching in K12 schools, and supporting College of Education faculty as Director of Technology. Stansberry is the founder and director of the Oklahoma State Student Media Festival, the recipient of the "Innovation in Teaching with Technology" award at OSU, a former K12 Teacher of the Year in Oklahoma, the Higher Education Representative to the Oklahoma Technology Association and past president of AECT's School Media and Technology division.

Christopher Stapleton

is the President of Simiosys, LLC a commercial simulation research company specializing in experiential entertainment, education and training applications. Known for his pioneering work in Mixed Reality Research as the Founding Director, of the University of Central Florida (UCF) Media Convergence Laboratory (MCL), where he has invented novel interfaces melting the boundaries between virtual, real and imaginary worlds. He is the 2009 General Co-Chair of the International Symposium of Mixed and Augmented Reality.

His company has collaborated with universities worldwide to create? Real World Laboratories? that bring together the academic, commercial and civic sectors to help define and refine the next generation of innovative products based on emerging technology, talent and techniques. He has 30 years of producing, designing, inventing and teaching entertainment from Broadway, feature films and television to acclaimed theme park attractions world wide. Mr. Stapleton received his Masters in Fine Arts in the design for film and theater at the Tisch School of the Arts at New York University.

Gretchen Thomas

is an instructor at the University of Georgia. She has worked as a classroom teacher and a system-level instructional technology coordinator. Currently, she teaches undergraduate courses on K-12 technology integration.

Index

A
accountability, 171
action figures, 63
activity theory, 107
AECT, 76, 221
alpha version, 160
analogies, 100
Animal Crossing Wild World, 80, 108, 123
APA, 195
art bible, 142
artificial intelligence, 84
assessment, 242, 250
autonomous, 170

B
Barbie, 60
behavioral learning approach, 140
beta version, 160

C
casual games, 241
catalyst, 61
character identification, 60, 65, 72
Code Red 911, 173
cognitive strategies, 100
collaborate, 236
collaboration, 170, 225
collateral learning, 227, 230, 248
community, 110
competition, 170
concept development, 137
concept development phase, 136
constructionism, 28, 29
constructivism, 111
constructivist, 29
constructivist approach, 140
constructivist learning environments, 111
context, 27
controversy, 194
COTS, 175, 201, 210
criterion referenced instruction, 192
cultural implications, 239
curriculum development, 244

D
dark side, 203, 213
deliberate practice, 192
Department of Health and Human Services, 206
design activities, 26
Dewey, John, 29
diegetic world, 28
differentiated instruction, 167
digital games, 224, 239
digital immigrant, 165
digital native, 165
disconnect, 234
division of labor, 111
document, 136
Dogs and Jackals, 59

E
educational game design, 223
educational games, 222
elaborative sequence, 95
ESRB, 196, 198, 207, 209
evolution, 245
expert reviews, 158

F
Family Entertainment Protection Act, 199
Fatallty, 191
feedback, 92
field tests, 159
field trials, 161
first principles of instruction, 102
flow theory, 32
formative evaluation, 158, 160
Froguts, 122

G
G.A.M.E., 175, 180
Gagné, Robert, 78
Gagné's nine events of instruction, 87, 192
game, 27, 49
game design, 77, 84, 133
game designers, 223
game mods, 198
game proposal, 136
game-based learning, 130
games generation, 166
game-to-classroom map, 177
GD process, 134
gender barrier, 63
gender based education, 68
gender bias, 62
gender differences, 58
gender divide, 64
gold version, 160
grounded design, 151
grounded instructional events, 155
grounded instructional strategies, 150

H
H.U.D., 85

258   Index

Hays Code, 209
hidden content, 226, 227, 249
high concept, 136
high stakes testing, 243
holistic learning environments, 242
homemade, 26

I
id model, 86
ID process, 134
identity, 50
immediate feedback, 192
inappropriate content, 249
instructional approach, 139, 140
instructional design, 79, 84, 130, 222
instructional designers, 223
instructional strategy, 131, 244
instructional technology, 79
Islamogames, 195
ISMF, 239
iterative cycle, 40

K
Kohlberg, Lawrence, 201

L
L.O.D., 85
Leap Frog, 172
learner analysis, 137
learner assessment methods, 146
learner control, 101
learning, 29
learning differences, 167
learning principles, 167
learning theory, 45
Leont'ev, Aleksei, 107
lesson study, 176
Luria, Alexander, 107

M
magic circle, 189
Magie, Lizzie, 60
Mancala, 59
Mario, 81, 123
marketing, 204
massive multiplayer online games, 206
mastery learning, 192
Math Blaster, 83, 112, 172
media rating systems, 207
mediating elements, 109
Merrill, David, 102
MMOG, 206
MMORPG, 175
mod, 28
modding, 28
Monopoly, 60

moral development, 200, 202
moral development theory, 201
morality, 212
motivation, 27, 180
Muzzy Lane, 237

N
negotiation, 30
nine events, 78

O
one-to-one evaluations, 159
open, 36, 38
organized course structure, 94

P
Papert, Seymour, 29
PBL, 116
pedagogy, 235
peer oriented, 67
performance, 91, 92, 243
Phoenix Wright Ace Attorney, 81, 82, 88, 118, 121
play, 60
player-versus-player, 207
point of view, 154
positive behavior intervention supports, 173
potentials, 245
PowerPoint, 26
Prensky, Marc, 166, 211
pre-production phase, 142
prior learning, 90
problem-based learning, 116, 170
production phase, 160
professional learning communities, 174
promotype, 136, 143
prototype, 39, 143

R
rating, 207
real-world problems, 104
Reigeluth, Charles, 94
Reigeluth's elaboration theory, 94
response to intervention, 173
retention, 93
role playing, 60, 64
roles, 170

S
Senet, 59
serious games, 234, 241
serious play, 31
shopping centers, 61
simulation, 69
situated learning, 49
situated learning matrix, 49, 52

small group evaluations, 159
social conditioning, 64
social domain theory, 200
social identity, 47
StageCast, 33
Star Games project, 176
Star Wars, 204
stereotypes, 63
stimulus, 90
student motivation, 28
Super Mario Bros, 80, 83, 104, 112
synthesis, 98

T
team-games-tournament, 68
technical design document, 142
Tet Offensive, 172
The Sims, 61
Trainien, 153

U
ultra-mobile personal computers, 176
understanding by design, 180

V
video game meals, 229
video game-based learning, 188
video games, 65
violence, 195, 226, 227, 249
violent video games, 194
virtual experience, 248
Vygotsky, Lev, 107

W
walnut shell game, 59
Walt Disney imagineer's ten commandments, 137
WebQuest, 30
within lesson sequences, 96

Printed in the United States of America